What the Flock!

What the Flock!

Raising kids, rearing animals and
other misadventures on our family farm

SALLY URWIN

THREAD

Published by Thread in 2022

An imprint of Storyfire Ltd.
Carmelite House
50 Victoria Embankment
London EC4Y 0DZ

www.thread-books.com

ISBN: 978-1-80314-377-4
eBook ISBN: 978-1-80314-376-7

The experiences described in this book are genuine but some
names and identifying characteristics have been changed
to protect the privacy of individuals.

For my very patient husband Steve and my gorgeous kids – I love you to the moon and back. Without you, this book would never have been written.

To my lovely mum and dad – thank you for always being there. With advice, help, regular cups of tea, deliveries of food, money (sorry about that) and buckets of love.

CHAPTER ONE

....

High Heels to Wellies

....

*Agriculture is our wisest pursuit, because it will, in the
end, contribute most to real wealth, good morals
and happiness.*

Thomas Jefferson

Twenty-five years ago, I was sitting in a toilet stall in a smart
accountancy firm, eating a snack pack of custard creams and
psyching myself up to deliver a presentation on the importance
of networking to some bored insolvency practitioners. I lived in
a minuscule apartment in a big city, had acrylic nails, very high
heels, a wardrobe of polyester suits from the petite section of
Dorothy Perkins and a permanently miserable expression.

I hated office life. Utterly despised it. I loathed the way it was
so strictly regimented, with everyone forced to sit in tiny cubicles
for eight hours each day, five days a week. Like the rest of my
colleagues, I did short bursts of productive work, watched the
clock and spent the rest of the time planning how to eke out the
pitiful remnants of my holiday allowance.

On the surface I was an engaged employee, enthusiastic about learning new developments in the field of accountancy marketing. Underneath, I had a drawer full of 'rescue remedies', drank far too much wine and I spent a *lot* of time sitting in the work canteen, eating bacon sandwiches and feverishly planning my escape.

In the office my polished veneer of self-assurance was only a few molecules thick and deep down I felt like a rabbit in the headlights, trying to give an impression of being ambitious and professional whilst longing to be back at home, snoozing peacefully under my duvet with Cyril the cat. (Note: Cyril was the best cat ever. I adopted him from a rescue centre. He was black with a white moustache and tail, liked cheese and onion crisps and I treated him like he was my baby. Sometimes I think if I hadn't found someone who wanted to marry me, I might still be in the same flat and dressing Cyril and his descendants up in baby clothes.)

It wasn't as if I was even interested in financial marketing, and it had certainly never been a burning ambition of mine to work in an office or work my way up the corporate ranks. After leaving Durham University with a history degree, I had absolutely no idea what to do. I've always been really impressed by young people who know exactly what career they want from an early age and are laser focused on the end goal of becoming a solicitor or a doctor or a teacher.

It had been drummed into me by the rather scary nuns at my all-girls school that women could do anything they wanted and should strive to achieve and be just as determined in their careers as any man. We were told that we needed to become high-flyers who worked just as hard and earned just as much money as any man, as well as remaining good Catholics and bringing up enormous, god-fearing families. Unfortunately for Sister Pauline, I had about as much ambition and ability to break through a glass ceiling as a damp sock.

By the age of twenty-one my relevant skills for the world of work consisted of the ability to fall asleep in lectures and drink two pints of cider and black without throwing up. In our last year at university, my college held regular careers seminars and each student was expected to attend at least once. They were held in the principal's study and were billed as 'Informal Networking Events' (which already sounds like something most sane people would prefer to avoid). One scorching June day I joined ten other twenty-one-year-olds in the overly warm study of Professor Ricketts (Emeritus in Medieval Archaeology, Principal of St Ovary College, Dunelm), nervously clutching a lukewarm glass of sherry.

Professor Ricketts was short and round with a long, grey beard and yellow horse teeth. He was our careers advisor as well as Chair of the college. He customarily spent his working day locked away in his dark study and tried very hard to avoid any members of the student body. Today he was dressed in a yellowing cricket jumper and tweed knickerbockers and had obviously partaken of a considerable amount of liquid refreshment before the networking event.

'And… have you considered staying for a Masters or even… um… a PhD?' he burbled, spilling his sherry on the carpet.

We all murmured vaguely in answer, longing for the chance to escape out of the French doors into the fresh air.

'Well…you have to decide what subject you're *most* interested in, oh my word indeed yes.' There was a pause as Prof Ricketts tipped back his glass and drained the dregs at the bottom. He peered owlishly at our young, upturned faces. 'There's no point deciding to study French Revolution politics whilst your real interest lies in, say, 1860 London brothels, oh gosh no.'

There was an uncomfortable silence as the professor made an attempt to pull himself together. He put down his glass and wobbled around on one heel to peer at the circle of students. 'So… young people. Yes. Tell me what you're going to do after you graduate.'

We stood in an awkward circle as each student told him what they were planning to do. 'I've been accepted by Geriatric and Mould Solicitors to train as a lawyer. They're number one in the world in coal mining law,' said one earnest-looking boy.

'I'm off to Grimsby for a gap year helping the impoverished and insane, and then I'm joining Flannel and Nutter LPC on a fast track to management,' piped up a smug girl with a Home Counties accent.

'I'm joining NASA as an integrational metals scientist to work on my antigravity rotational transporter,' said an intense-looking girl, smiling proudly around the circle.

I realised that everyone was either joining a world-famous law firm or embarking on a serious academic career in something unintelligible. My face grew hot, and I edged behind the boy in front of me.

'And you…' Prof Ricketts slurred, noticing me for the first time. His eyebrows went up. 'Gosh, aren't you… very small?

Ahaha.' I glared at him whilst the other students tittered behind me. 'What do *you* want to do with your degree?'

My mind went blank, and in a very tiny voice I said, 'I'd quite like to live in the country and keep chickens.' My comment dropped like a heavy stone into a pool of sudden silence.

Prof Ricketts opened and closed his mouth a few times. The rest of the students snickered behind me as my face heated up in mortification. Prof Ricketts – bless his ancient, pockmarked soul – turned to the next person as if I'd given a perfectly sensible answer.

I spent the next few hours sulking in the college bar trying to drown the memory of my reply with cheap lager. In essence, what I had said to the professor *was* absolutely true: instead of an ambitious career, I envisaged myself making a home like my own mum and her mum before her, bringing up a family in the countryside, maybe with a small part-time job and even enjoying the quiet excitement of the odd coffee morning. I dreamt about a peaceful, quiet, rural type of life, with horses, chickens and dogs and a warm kitchen.

I was young and very naïve.

I also suffered from an intractable case of social anxiety. I had always been very shy, and I found it an immense effort to speak to people I didn't know. Especially those who were a lot bigger than me (which, since I was such a tiny dot, was almost everyone). I was the kid who cried every day at school, hid behind my mum's legs and often had tummy pains at the thought of leaving the house. Nowadays there is help for children who have this intense anxiety and shyness, but in the mid-1990s my terrors were dismissed as 'just a phase' and 'something she'll grow out of'. (Note: My school rounded off my first report with the withering comment: 'Sally has the ability to find problems where there are none.')

Many of my friends had been snapped up in their last year as a student during the terrifying 'milk round', when large, impressive national companies toured the universities to drink sherry and recruit the most promising undergraduates as junior analysts and baby executives. I had been too shy to join in and massively intimidated by the whole idea of being on a 'fast track' to management. So, in July 1995 I quietly moved from my scruffy student digs to my mum and dad's back home, clutching my stripy duvet cover and posters of Depeche Mode, and vaguely hoping that some kind of job would just fall into my lap.

As well as the occasional desultory job application, I slipped back into a peaceful routine: walking in the Cheviot Hills, visiting hill forts and stone circles with my dad and inspecting the cup and ring marks that were scattered on the sides of hills such as Simonside or Lordenshaw Crags. I'd been brought up in a tall, Victorian terraced house that overlooked the North Sea, and as a young girl, the thick sea fogs and high winds were in high contrast to the bucolic peacefulness of my grandmother's house in Felton, a small rural village in Northumberland. When I was younger, I spent many happy days staying in 'Granny Felton's' tiny eighteenth-century cottage, and her affinity with the farming landscape, the wildlife and the birds that visited her garden taught me about the countryside; even now the 'chuck, chuck' of the jackdaws and smell of honeysuckle and roses immediately transport me back to those peaceful, happy country days. Spending holidays in Felton left me with an enduring love for the rolling hills and valleys of Northumberland, and after university, I spent many contented hours sitting in the hills on the sheep-cropped turf, basking in warm sunshine, drinking a flask of tea and watching the shadows of clouds move across the valleys, listening to my dad pointing out the names of the little farms and cottages far below us.

After a few lacklustre job interviews I reluctantly started work as an administration assistant in the marketing department of a local firm of solicitors. With a degree I was over-qualified for the post, but I settled into the routine of filing and typing and getting to grips with the emerging 'world wide web'. Over the next three years I moved gradually up the ranks, learning how to hide my shyness by skimming a veneer of self-assurance over the nerves. Looking back, I realise now that working in a busy, loud office and pretending to be the assertive, extrovert personality that the job demanded drained every last ounce of my energy. So, to recharge and soothe my frazzled nervous system I would drive up to the Northumbrian countryside every Saturday morning, trailing for hours around English Heritage castles and National Trust houses with my long-suffering parents, who must have been secretly wishing I'd find a social life of my own.

I felt claustrophobic and hemmed in at the office, with all its rules and regulations, and trussed up in my tight suits and high heels. Taking the time at the weekends to pull on my walking boots and yomp through miles of heathery moorland or climb a windswept hill in the Cheviots gave me a much-needed sense of balance and space to think about the week ahead. I've always been fascinated with the history of this part of England, and I found that learning about the struggles of previous Northumbrian inhabitants put my own anxieties into sharp perspective. Poking around old ruins and decaying stone forts and tombs of long-dead inhabitants reinforced the truth of the self-help mantra 'this too shall pass'.

Eventually, I gained enough experience and social polish to apply for a job as a marketing manager for a big, national accountancy firm. After a gruelling round of interviews and presentations I was offered the role.

I can't overstate how big a deal this job was to me at this stage in my life. After university my increasing social anxiety, intrusive thoughts of harming myself and general feeling of doom had eventually forced me to tearfully stagger to my GP and, whilst sobbing loudly, pour out a garbled litany of worries surrounding my low self-esteem and crippling anxiety. My doctor offered me a tissue, quickly enrolled me on a course of therapy and prescribed a big, fat daily dose of an antidepressant called Seroxat. With his sympathetic help, a young therapist and the tablets, I slowly improved, and I managed to overcome the terror of speaking to people. The horrible thoughts began to recede, and life, although still difficult, became less of a daily struggle.

When I was given my own business cards, expenses card and laptop, it felt like I was finally fulfilling the expectations that were created at school and university.

This was what ambitious women did! Sod my wish for coffee mornings and keeping chickens! Real women worked hard, got good A levels, went to a good university and finally had a stellar corporate career, with no messing about. I *needed* to succeed at this job.

I realise I was trying to sledgehammer a round peg into a square hole. The office was desperately boring, and the work didn't interest me in one iota. I did like the income and wearing a nice suit, but I found pretending to be enthused about financial marketing quite exhausting. On stressful days I used to go up and down in the office lifts, admiring my outfit in the mirror and buying myself a little time before I had to strap on a cheerful face and go back into my office cubicle. (If you kept your finger on the 'close doors' button in the lift no one else could get on and I could have a few minutes of private deep breathing.)

One week into my new job, my boss invited me down to London to meet the Contingency Planning and Insolvency Services Department and to join the firm's annual awards dinner. On the trip from Newcastle to King's Cross I tried very hard to fight down the waves of anxiety, helped by a handful of beta blockers and a liberal application of extra-strength deodorant. I hailed a black cab and sat in the back whispering cheesy confidence mantras to myself until the driver pulled up outside an enormous office with black-tinted windows.

The reception desk was an intimidating bank of black leather, and behind it, glossy young women with matching suits tapped away at computers as I threw back my shoulders and squeakily asked for the Head of Marketing.

I was ushered into a beige meeting room and asked to introduce myself to a motley group of financial managers and accountants. This was the team that I was going to promote, who had the unenviable job of winding up bankruptcies and unravelling dying businesses. Everyone was male, white, over 40 and had that pale doughy appearance arising from too many motorway coffees and badly digested fried breakfasts. They drank cups of coffee and texted on their phones as I fumbled to the top of the table. I had prepared a little speech, and I launched into an enthusiastic spiel about financial marketing with lots of nervous hand movements and the occasional joke.

Everyone stared at me after my little talk. My manager stood up.

'Does anyone have any questions?'

There was an awkward silence.

'Right, well if that's the case I'll ask Sally here to step—'

'You're a proper Geordie!' called a man at the bottom of the table.

'Well, in fact I don't really think I do have a strong accent, and actually my parents…'

'Say "Howay the lads",' he chortled, with an attempt at a terrible Newcastle accent.

I discovered that when surrounded by plummy accents, I did sound *very* northern indeed, and I had many conversations about the nightlife in Newcastle, the fact that no one wore a coat and whether I knew a bloke called Alan that had moved up and now lived in Byker.

Being a young woman, the day also reintroduced me to the concept of 'mansplaining', in which older, dull men explained things to me that I already knew. Mostly about the history of the North East, or the best way to drive down the A1 or even, at one point, the proper height of heel I should be wearing on my shoes. If mansplaining ever became an Olympic sport, then that group of accountants would win a gold medal every single time.

My boss was 30 years old, boundlessly ambitious and completely terrifying. 'Oh, you're so confident!' she cooed after my first meeting. 'You've obviously found your niche working in marketing.'

I smiled weakly and jerkily pushed my hands into my too-long sleeves. 'Yes, I love the world of insolvency marketing, and I'm so excited to be here,' I said woodenly. Fortunately, she smilingly ushered me into a darkened auditorium for the main event of the day, a chance for the top management team to lecture us about the next year's key performance indicators. It was a relief to let my face drop from its fixed rictus grin, close my eyes and try some more deep breathing.

That evening there was an event billed as a 'networking cocktail party' where I spent a lot of time listening to middle-aged accountants discuss their new company cars or the best time to hit the M1 on a Monday morning. I stared longingly at the canapes and took big gulps of my plastic cup of warm, white wine.

After a while, a big screen was rolled onto the stage, the lights dimmed, and we all took our places at our dinner tables as a fanfare of trumpets announced that we were going to sing the Company Song. An animated sequence started up on the screen, with a rubber ball bouncing along the tops of cartoon lyrics so that everyone could join in. With a happy roar my table started belting out a tub-thumping chorus to the tune of 'Rule Britannia'. I'd love to tell you the actual lyrics, but I think my brain has erased them for my own sanity. I'm sure there was an accordion accompaniment involved. There was certainly a lot of shouts of 'team spirit and winning goals'. After the last rousing chords had died away, we were called up onto the stage and presented with a tiny enamel badge. Some blokes on the same table were given big gold badges, and they acted as if they'd won the lottery with lots of whoops and back slappings.

Then the booze appeared…

Eventually I dragged myself away from the drunken hilarity and staggered back to my hotel. Sitting on the bed I eased off my shoes and massaged the balls of my feet. The realisation that I'd have to get up and do this all again tomorrow was clanging urgently around my brain. And I'd have to travel to different offices around the UK and repeat the same round of meetings and pretend that I cared about accountancy firms, insolvency departments and middle-aged number crunchers called Brian.

At that moment, I felt hollow inside. How the hell was I going to cope with it all? How would I keep up the exhausting pretence of being someone I wasn't? Where was the joy or excitement? I imagined myself ten years in the future, still slaving away with twenty days of holiday a year (minus bank holidays) and counting down the months until I could go on a fortnightly annual holiday. I suddenly felt completely and hopelessly trapped.

I had another six months of endless travelling, gruelling meetings and terrifying presentations, until I came down with a proper gnarly dose of flu and then just… never really recovered. I'd been experiencing a vague 'glandular fever'-type illness since my second year at university, and sometimes, especially during stressful times, it would flare into a temperature, sore throat and deep fatigue, leaving me exhausted and listless. After this bout of 'flu' I was sleeping fifteen hours a night and needing naps every couple of hours during the day. I was hollow-eyed with tiredness, and each day felt like I was trying to drag myself forward through a deep vat of treacle. My GP signed me off work for two weeks. The symptoms started to become worse, and the weeks stretched into months.

Looking back, it was obvious that the doctors just didn't know what to do with me. Endless blood tests came back negative for Lupus, Lyme disease and glandular fever. Eventually, more in frustration than with any conviction, I was diagnosed with ME. In those days it was known as the 'yuppie flu', and the diagnosis was often met with an eyebrow raise and a disparaging comment about being a 'work shirker' or 'downright lazy'. (Note: ME is myalgic encephalomyelitis/chronic fatigue syndrome, a disabling and complex illness with the primary symptom of overwhelming fatigue that is not improved by rest. There is no known cure.)

I think anyone who has dealt with an invisible disease knows the deep frustration and internalised shame that accompanies the symptoms, and I became very good at gaslighting myself. Maybe it *was* just all in my head, and maybe I *was* just a hypochondriac. Or, as a charming relative once put it, just needed to 'get off my backside and do more exercise?'

I tried everything – acupuncture; antibiotics; having my tonsils taken out; therapy; pacing; antidepressants; meditation and mindfulness – and found that absolutely nothing made any difference to the crushing tiredness and weakness. After a lot of wasted appointments and a pile of money, I finally came to the

freeing conclusion that all the so called 'treatments' were absolute
bollocks. All the supplements and therapies that were touted as the
cures for this newly recognised syndrome hardly helped at all. All
I knew was that each flare-up just lasted as long as it lasted, and
the only thing I could do was to ride it out and rest through the
worst of the exhaustion. Trying to push myself and fight against
it just made everything worse.

The illness meant that I had to sell my little flat and (dragging
Cyril the cat along with me) move back in with my parents, and
with their help and love I managed to get back to my job and
cope for another twelve months. To be fair, the company did
their best to make allowances; I stopped travelling as much, and
eventually I began to feel a bit better and could even enjoy the
occasional night out.

By this time I'd had a few relationships, but they had never
lasted more than a few months. I'd been single for a little while
and hadn't had much luck in finding anyone special in the clubs
and pubs of Newcastle. In the early 2000s internet dating was in
its infancy and there were no apps like Tinder or Plenty of Fish.
The only UK site in those early days was called Dating Direct,
and it was a big clunky database with an awkward web interface
which you could search through to find someone you wanted
to meet. Using a dating website carried a stigma in those early
internet days, but I was in my late twenties, had a string of failed
relationships behind me, and I wanted to find someone who
shared my own values and beliefs.

Once you'd created your Dating Direct account there was a
page where you could click on a long list of different personal

qualities that you wanted in your next partner. It was a comprehensive list and covered such important virtues as political leanings, height, salary range, confidence levels and even hair colour. It was a lot of fun, and I clicked the little buttons to instruct the algorithm to search for a man under 5 foot 7; between thirty and forty; who lived or worked in a rural area; who had all his own teeth and liked animals. And the very first profile I saw was Steve, a sheep farmer, who was grinning out nervously from his photo and wearing the most terrible oatmeal rollneck jumper.

Our very first date was at The Rat Inn near Hexham. I'd briefed a friend to ring me halfway through the evening in case Steve was peculiar or showed any obvious serial killer personality traits, but apart from shaky nerves on both sides we got on like a house on fire.

Regrettably, I'd also asked my dad to come and pick me up at the end of the evening, and instead of staying inside his car to wait like any normal person, he strode into the bar and, as I looked on in horror, firmly shook hands with a startled Steve.

'Good evening,' he boomed. 'Richard Dixon, retired engineer.' Dad spoke with the happy confidence of an important speaker introducing himself at a company dinner or maybe to the local Women's Institute. I glared at him in embarrassment, managed to extricate him from the pub and back into the car, and was thankful to discover that Steve didn't seem to be put off by meeting my family at such an early stage in the relationship.

Our second date was in the middle of lambing time. I had absolutely no idea what that entailed or what I would be expected to do, but Steve told me to wear a pair of wellies, bring a flask of tea and gave me careful directions to his farm. I still managed to become hopelessly lost in the maze of tiny, winding countryside lanes, and an hour late, I nervously nosed my tiny car through big farm gates and into a very dusty yard. The weather was glorious with bright, hot sunshine, and as I pulled up I saw a very tired-looking Steve trying to shoo a flock of fluffy shapes into the sheep pens. Unbeknownst to me Steve had been up all night and was running solely on nerves, caffeine and chocolate, but he managed a big smile when I got out of the car wearing brand new, very clean wellies and a bright yellow anorak.

I waved the flask of tea and he gratefully settled down on a straw bale for a cuppa as I looked interestedly around the pens of different sheep.

'It's all a bit messy and noisy,' warned Steve as he pointed out a ewe in full lambing mode who was pawing at the straw in her pen, turning in small circles and making faces at us over the bars.

I was absolutely fascinated and asked a million questions, trying to get a handle on why some sheep were inside tiny pens and some were in the bigger shed and how much help he had (zero) and what he did if a lamb wasn't well.

'Come on, I'll show you,' said Steve, putting a stop to my torrent of queries. He hopped off the straw bale and helped me open the gate into the main lambing pen, ushering me to the side of a ewe who was lying down in the straw and seemed to be pushing like mad. My eyes were like saucers. I'd never seen a sheep give birth. Hell, I'd never seen *anything* give birth. I was twenty-nine years old and my expertise was writing press releases and organising cheese and wine networking seminars. I stood to one side, awkwardly holding a bottle of lambing gel, whilst Steve flipped the ewe onto her side, reached around to her backside and gently drew out a pair of tiny, wet hooves.

'Oh my god, that's a lamb!' I breathed as Steve gave a huff of laughter.

'Aye, I'll just need to give it a quick pull, as the shoulders are wedged a bit.' With that he pulled the tiny feet, and after a few tense seconds a floppy, wet lamb slid out of the ewe and onto the straw. It lay for a minute, sides rising and falling with the first breaths, as Steve wiped the goop from its nose and gave it a rub down with a handful of straw. The little baby shook its head and gave a high-pitched bleat, which was immediately answered by the ewe with a rumbly mothering chuckle.

'That's amazing,' I said. 'It's a real, live lamb.' I knelt down in the straw and stroked the little, wet ears and muzzle, feeling the flick of a warm, wet newborn tongue over my fingers. And embarrassingly and without any warning, I burst into tears.

Steve sat back on his haunches, rubbed his nose and looked at me with his eyebrows raised.

'Oh god, I'm sorry,' I snuffled, wiping my nose on my sleeve, 'it's just... just a miracle and I've never seen it, and it's so cute and the mum is so clever and...'

I still cringe at my outburst, even after all these years, but Steve was very kind, offered me his slightly grubby hanky and, after moving mum and baby into a pen, gave me another cup of tea and a Mars Bar to recover.

He was too busy to sit down for long and rushed off to collect ewes and lambs from the paddock to move them into a field. I started to give him a hand and he quickly showed me how to safely pick up a lamb and move them into a pen. (You hoick the baby up by both front feet in one hand, make sure that the ewe can still sniff and see her lamb, and then move slowly step by step into the pen, hoping that the mother will follow. Sometimes she does and sometimes she doesn't…)

'Right,' he said, pointing at a very young pair of twins and their proud new mother. 'Go and pick up those two lambs and move the family into that top pen.'

I wandered up to the newly born pair and spent a few minutes trying to grab both front hooves in one hand. The legs were slippery with amniotic fluid, and the lambs were surprisingly heavy. I discovered that I had to lift each lamb up to shoulder height so that their back legs didn't drag on the straw. Whilst I was fussing around, the mother decided she was hungry and disappeared into the crowd of sheep munching away at the sheep feeder. I now had two heavy lambs in my hands but no following ewe. Instead, three other sheep wandered over and started trying to lick the soggy lambs, all making that 'mothering chuckle' noise I'd heard previously. I chose the biggest ewe and tried to shoo off the others, turning my back so I could herd the new little family into a pen. My arms were aching with the weight of the lambs as they dangled helplessly from my hands, and the three competing ewes clustered round, banging into my hips and knees as they jockeyed for position, all trying to nuzzle into the now shouting babies.

'Not that bloody ewe. *THAT BLOODY EWE!*' bellowed Steve from across the pen.

What? What the hell did he mean? Which ewe? I had three of the buggers all trying to adopt the lambs as their own.

At this point I was jammed into the corner of an individual pen with the two babies across my lap and three huge Texel ewes crammed into the tiny space, butting my legs and knees as they tried to reach the bellowing lambs.

'Help!' I screeched, taking off my bobble hat and whacking one aggressive mother over her head.

'Oh for god's sake,' Steve muttered as he climbed into the pen, easily shoving the three enormous sheep back through the gate and grabbing the real mother from the silage, pulling her away from the sheep feeder and into the pen and forcing her head down to the lambs where she immediately began to lick them dry.

He grabbed me under the elbow and helped me up to my feet. I stood shakily, trying to dust off the goop and straw from my grubby-looking anorak.

'What happened?' I said. 'I couldn't work out which one was the mother. How do you work it out?'

Steve had been up twenty-six hours without any sleep, and I could see him swallow convulsively as he tried very hard not to make a sarcastic comment.

'You look at their backsides,' he said carefully, 'to see which one is covered in amniotic fluid. They all wanted to lick the lambs as some are very close to giving birth to their own babies and their hormones make them confused.'

'Ahhhhhhh!' I replied, light dawning. 'Well, you should have told me that to begin with, and I wouldn't have made a mistake.'

I climbed out of the pen and started helping again, filling buckets and water troughs, trudging through the ankle-deep straw to feed and water the animals.

Years later Steve told me that it was at that exact moment, watching me scrub out filthy water buckets, that he realised that our very new relationship might develop into something more serious.

I recently asked him about this sudden flash of realisation.

'But didn't you want to go out with a farmer's daughter?' I asked. 'Someone who knew what they were doing and didn't have to be told what to do every step of the way?'

'Nope,' he replied, rubbing his hand up and down the back of his head. 'It's not that I minded explaining stuff, but the thing that I liked was that you didn't get upset that I'd shouted or said you did something wrong. You just picked yourself up and carried on.' He thought for a few seconds and added, 'It's like breeding sheep anyway. It's important to find someone from outside the area to introduce a bit of hybrid vigour into the family.' He's so very romantic.

It was painfully obvious that I knew absolutely nothing about farming, apart from reading a few James Herriot books, but as our relationship developed I began to gain more of an understanding of how the land and the animals slotted together. Our subsequent dates were all on the farm and I saw different snippets of the farming world, little excerpts of birth to death and everything in between. I was hooked and I still am, although my knowledge has improved a little since those very first days.

After a year of dating, Steve asked me to move into his tiny cottage, and along with a slightly confused Cyril the cat, we settled down together to life on the farm. After a couple of months of commuting backwards and forwards to the city centre, I finally made the momentous decision to hand in my notice to my boss. Is there anything better than handing in your notice to a job that you hate?! My stress levels almost immediately reduced, and whilst the tiredness and other symptoms were still there (and would always be around in some capacity) my sleep improved, and I had more room to plan and start thinking about what I really wanted out of life.

That Christmas, in front of a roaring fire and a snoozing Cyril, Steve went down on one knee and produced a diamond solitaire ring and asked me to marry him. I said yes with no hesitation at all and (again) burst into tears.

*

After Steve and I were engaged, I still had rose-tinted expectations of living on a farm.

I imagined myself wafting around in long, floral dresses with well-groomed hair, being followed by a bevy of adorable, rosy-cheeked children. We would be sitting on a woollen rug at the edge of a picturesque hay field, eating a delicious and nutritious homemade picnic whilst my cheerful husband brought home a bounteous harvest, only pausing for a happy family walk among the cut hay. Probably with a cute white pony in the background and an obedient collie dog sitting by my feet.

Or, in shorthand, like one of the irritatingly smug adverts from a Boden catalogue.

The reality is a little different here at High House, the farm Steve and I run together.

I'm usually wearing leggings and a huge, unattractive jumper with a sheep poo mark down the front, and I've forgotten to brush my hair. My kids are bickering about being dragged outside when all they want to do is play on the Xbox, and the picnic is a few elderly sausage rolls and wizened satsumas thrown together in a tremendous hurry. Steve can be guaranteed to be in a filthy mood about the weather or the non-appearance of the combine harvester, and instead of a rug, we're perching on a few slimy feed bags and being pestered by wasps, whilst I'm bellowing at Mavis the collie to stop chasing our grumpy pony in the field next door.

It *can* be idyllic. When it's not blowing a 'hoolie' (howling gale), the sun is shining, nothing important has died and the tractor hasn't sprung a leak, or we're not afflicted by any of the other hundreds of misadventures that can befall a farmer... well, *then* life is pretty wonderful. But good scenery and plenty of outdoor space don't always mitigate the fact that it's a lot of hard work for very little material reward. We are often living hand to mouth, and we are reliant on so many things out of our control (such as decent weather and good lamb prices) that we walk around in a constant fog of low-grade anxiety about the future of the farm.

I meet a lot of people at High House. We have a new pig shed, sheep, a tearoom, a wedding venue and a brewery operating in the main part of our farm buildings, and the paddock is now rented out to a glamping company. Many visitors enjoy watching us work and chatting to our various animals, and I can guarantee that once a day someone will gaze expansively across our rolling fields, sigh deeply and say, 'It must be an *amazing* life living here. All this space. I'd love it to be able to work outside all day.'

Then I point out that we work ridiculous hours for little or no money and we're often outside in the freezing wind, snow and rain, sometimes with very little sleep and with no chance of a rest or the option of a holiday.

This doesn't mean that I don't appreciate our gorgeous scenery and the slower pace of life. I am enormously grateful that I can bring up my children in such glorious surroundings. I realise that I am hugely lucky to be away from the nine-to-five, corporate grind. But living in the country is not always the twinkly rural idyll inhabited by happy, apple-cheeked farmers that you might expect…

But the lifestyle definitely suits me better. Farming can still be a cause for anxiety, but the pace of each day is much slower. My kids have acres of space to play in and enjoy all the benefits (and frustrations) of a rural childhood. They even occasionally work alongside us, especially during lambing.

Ultimately, I've found the kind of life that I craved and I needed: watching the seasons change and immersing myself in the slow patterns of a farming year. And, as a bonus, I've never had to sing a corporate song ever again.

After I left my job and joined Steve on the farm, I would occasionally receive phone calls from recruitment agencies asking if I was interested in returning to the 'fast-paced and exciting world of financial marketing'. I would stop whatever sheep-based chore I was doing, mobile phone clamped between my ear and hunched shoulder, as I listened to the spiel from the agent listing all the benefits I could expect to receive in whatever accountancy firm they were talking about. No matter how many days holiday allowance or amount of pension contribution they offered, the answer was always 'no'. I would politely decline, push the phone back into my boiler suit and, with a sense of relief, turn back to trimming feet or mucking out a stall. Eventually they stopped calling. Each day on the farm was different, and although some of the work was backbreaking or muddy or the weather was cold and wet, it was still three hundred per cent better than sitting in a dusty, artificially lit office, attending the same deadly boring work meetings and trying not to panic over ever-tightening deadlines.

My first visits to the farm opened my eyes to the joys of working in the open air, the experience of being outside in all weathers and being able to choose (to some extent) what I wanted to do each day. The first few weeks also taught me the basics of being around sheep and cattle, how to move quietly and confidently and the best place to stand to avoid being squashed against the side of the shed or run over in their field. From my very first date with Steve, I was full of enthusiasm and endless questions and ready to plunge headfirst into a brand-new life.

CHAPTER TWO

Far from the Mooing Herd

When life seems one too many for you,
Go and look at a Cow.
When the future's black and the outlook's blue,
Go and look at a Cow.

From 'Ode to a Cow',
The Old Farmer's Almanac (1936)

When I first met Steve he kept a small herd of beef cattle. I had never seen cattle up close. I'd seen distant groups of vague cow-shaped blobs in fields, but I hadn't been within touching distance of a real, live animal.

Sensing my enthusiasm, Steve took me to meet his herd after one of our very first dates. (Ah, the romance!)

I remember the feeling of excitement as we drove up to the farmyard, and I first caught a glimpse of the cattle shed looming out of the frosty dark. I lowered myself down from Steve's rickety Land Rover onto slippery cobbles slick with ankle-deep, icy mud. I was wearing a pair of shiny, black, leather boots, with at least a four-inch heel, and I slithered over the frozen ground, hanging onto Steve's arm, until I reached the dry concrete in front of the shed.

I'd never been in such complete darkness. I tipped back my head, cheeks sparkling with cold, and took in my first sight of the Milky Way with millions of pinprick points of light dusting a velvet navy sky.

The smell of a working cattle farm is thick and heady: so pungent and rich you can almost taste it on your tongue. Steve flicked on the shed lights. Twenty cows and calves stood there in steamy warmth, heads raised and silhouetted in a pool of yellow light. All was still and silent until a large, black cow at the back of the shed pawed the straw at her feet and gave a low moo of welcome, whilst her calf, ignoring us completely, nosed underneath her belly. The herd was split into two by a concrete walkway that rose up from a gentle ramp to run the full length of the shed, and on each side bulky metal bars faced each other, with slanted gaps so that the cows could push their heads through to eat. The front of the shed was covered by two enormous pitted and scratched steel-plate doors that were higher than my head.

Steve marched up the walkway and felt in his pocket for a penknife. He slit a black-wrapped bale of silage and started tossing wedges of hay onto the central ramp, and the cows ambled to the front of the pen and pushed their heads through the bars. I watched as they used their long, pink tongues to carefully curl around the dried grass and jerk a wedge into their mouths. The silage had a dense, fruity aroma, just like a banana cake. (I would quickly learn that although *fresh* silage smells wonderful, after an hour or two it begins to assume the distinct whiff of rancid cat pee.)

I leant over the top pen and stretched out a hand to touch the nearest cow's face. She was much bigger than I expected, with a curly, ginger hide and long, white lashes fringing a round, black eye. But before I could touch her, the cow snorted and spun sharply away towards the back of the shed.

Steve smiled and stepped up beside me. 'They don't like sudden movements,' he said, 'especially if they don't know you very well.'

It struck me how bulky and heavy each cow looked up close, with square-shaped bodies and angular hook-boned hips. Unlike the elegant, flowing lines of a horse's head, neck and back, each cow had a large, heavy head, a short, powerful neck and a stocky back that ended in a thin, tufty tail. They were skittish and ungainly and pushed roughly against each other as they reached for the slices of silage.

One cow stood out from the rest, with an auburn coat that shaded to ginger brown on her legs and tail.

'What's her name?' I asked.

'Red Cow,' said Steve, concentrating on pulling thick slices of aromatic grass from the bale and dividing them up between the herd. 'Not very imaginative, I know. She's really friendly. She's my favourite.'

Red Cow pushed her wedge-shaped head up against the railings and peered at me out of one eye. Steve showed me how to scratch around the base of her fluffy ears, and she bobbed her head up and down in delight as I worked my fingers into the warmth of her coat.

I'd never worked or been close to large animals, and after that first meeting I took every chance I could to help with the cows, shadowing Steve as he moved around the herd, learning how to look after them and gain their trust. I found real joy in being with the animals, learning their quirks and working hard alongside Steve in the cold air. At this point, I was still working at my marketing job in Newcastle, and I spent so much time with the herd that returning to my desk job on a Monday morning felt jarring, the grey office cubicle and Excel spreadsheets feeling claustrophobic and restrictive compared to being on the farm in the fresh air with my warm, furry-coated friends.

After a few months, moving into Steve's little, one-bedroomed farm cottage seemed like the next logical step, so one morning I packed up my tiny car with all my belongings, including a bewildered Cyril the cat, fifteen cardboard boxes of paperback books and a suitcase of clothes, and drove down the twisting roads to set up house on his farm. Settling into life together was a definite learning curve, both for Steve and myself. His sparse bachelor pad was soon filled with endless paperbacks, shoes and cat toys, and his bathroom shelves groaned with a collection of floral-scented shower washes and face masks. On the other hand, I had to become used to the constant muddy boots, early mornings and dark nights and learn how to navigate around the farm, even in the pitch black, without the benefit of orange sodium streetlights. The most noticeable difference was the peace and quiet. On the farm there was no constant low drone of traffic noise, aeroplanes or the shouting of noisy kids and passers-by. Instead, the only things I could hear were the occasional bleat of a sheep, the rumble of a quad bike or the natural sounds of the wind and the rain. It was absolute heaven.

I soon realised that farming life at High House had always followed a very strict timetable, and I was surprised to find out that Steve structured his days as inflexibly as any army barracks. Steve would get up at six o'clock on the dot, have his 'first breakfast' of tea and toast and then go out to work. He would come into the house for his 'second breakfast' of a bacon sandwich at half past ten, whilst lunch was at half past twelve (and not a minute later). An afternoon cup of coffee was drunk at three o'clock and dinner was on the table at six o'clock. I could set my watch by his routine.

After a while I broached the subject of why he organised himself in such a regimented manner.

'I'm not really sure,' said Steve, wrinkling his brow. 'It's just what my dad did. And I suppose I'm just doing the same.'

'What would happen,' I asked mischievously, 'if you had lunch at one o'clock instead of twelve thirty?'

Steve screwed up his face and laughed. 'God, I don't know. It doesn't feel right but I've no idea why.' After that, he tried to relax his routine a little, but if I wasn't around to distract him he would snap right back into the 10.30 a.m./12.30 p.m./3.00 p.m. timetable. I suppose generations of farmers from Steve's family had stuck to the same daily structure, and by this time it was ingrained as an ancestral memory, a way of working that carefully spaced hard, physical tasks with hot meals and reviving cups of tea.

In those days we had more cattle than sheep, and our cows were Simmental crosses: good beef animals with enough muscling and breadth to produce a healthy calf each year. The herd was a mix of older cows, who had years of calving behind them, and younger heifers, some of whom were in calf for the first time. In the early 2000s a single 'bulling heifer' (a young cow that had not yet been put in calf) cost around nine hundred pounds to buy from our local mart, and once bought they stayed at High House for another seven to eight years, producing and suckling a yearly calf.

In our 'suckler herd' the calves were kept with their mothers for ten months, and then – after weaning – they spent another eighteen months on our farm in smaller groups on the grass fields, slowly building up enough muscle and condition to be sold – back at the mart again – either for prime 'grass-fed' beef or as the basis of new breeding herds. In those days the selling price for a young bullock or heifer was around seven to eight hundred pounds – a fair return on our investment.

Life on High House Farm closely followed the natural yearly cycle of a cow's life: from mating with the bull, through pregnancy and the birth of a calf, suckling the calf until it was ten months old, weaning and then building her condition so that she was fit enough to be mated once again. A cow would not stand to be mated by a bull until she was ready, and we kept a close eye on the health of the older animals. Some of our more mature beasts had a propensity to suffer from mastitis (a painful swelling and infection of their udders) which sometimes damaged their ability to produce enough milk. Eventually a cow that was too old to produce a calf was either retired or, more likely, sent to the 'cull cow' sale.

Unlike the dairy industry, where calves are removed at an early age to maximise milk production, our babies were fed by their mothers until they were at a natural age of weaning. At ten months most of them had stopped drinking milk altogether, although there were always a couple who still tried their best to suckle, dropping to their knees to reach under their poor mother's belly to headbutt her udder so she would 'let down' her milk. I loved the fact that these calves then spent another two years on the farm in small, sociable herds, eating our grass, slowly putting on muscle without any need for growth hormones or other artificial additives. It took a long time to build the quality 'grass-fed' beef that the market demanded, and we prided ourselves on our very high-quality welfare standards, from birth to slaughter.

All the cows and calves spent the winter in the farm sheds, sheltering from the cold and rain in a roomy pen with a deep straw bed. Every few weeks, Steve would open the doors and gently shoo the cows into the farmyard whilst he dug out the thick layers of muck and straw with his tractor – dropping huge clods as he went – then carrying the manure down the road to the midden in the bottom field. Mucking out could only be done on a frosty day so the ground would be hard enough to bear the weight of the tractor and trailer without it becoming stuck in the mud. Everyone knew exactly what we were doing, as the overpowering, cloying smell of cows would hang in the air, the smell so strong that it permeated the house, giving *all* our meals a faint spice of cow muck.

*

In late spring the cattle started to look very pregnant, with rounded bellies and filling udders. In good weather, days were spent standing nose to tail in the long grass, flicking away flies with their tufted tails or lying flat out in the sunshine and grazing peacefully in a tight-knit herd. Bully (the imaginatively named bull) was put in a different field with our small herd of young bullocks, and they had a peaceful summer snoozing in the nettle patch with the occasional excitement of following a sweaty-looking rambler who'd taken a wrong turn.

The pregnant cows and young heifers became very tame, rushing up to our garden gate when they heard Steve cutting our lawn and waiting patiently for a cattle delicacy – a bucket of grass clippings. They were also rather fond of the occasional gift of a turnip and used to wait eagerly every morning to see if I'd found anything wizened in the vegetable rack.

One morning, when my daughter Lucy was barely more than a toddler, I looked out of the kitchen window and found her balancing precariously on a bucket, carefully feeding a punnet of (my expensive!) blueberries, one by one, to a very appreciative Red Cow. Red Cow and Lucy used to spend happy moments by the gate together, Lucy sitting on the top rung and singing her latest made-up song into Red Cow's fluffy ear. I had to keep my blueberries on a top shelf in the fridge as otherwise they'd go missing, and I'd have nothing to put on my breakfast yoghurt.

Lop Ear was another of our favourites. She was a young cow of a speckledy roan colour and grey-blue cast to her coat. The grey faded out to white around her muzzle and eyes, giving the distinct impression she was wearing spectacles. She had short, stumpy black legs and dainty hooves and an ear that curved down towards her face, rather than sticking out horizontally like most cows.

Lop Ear (again imaginatively named) was near the bottom of the herd pecking order and kept herself towards the back of the shed, trailing along behind the line of grazing cows when she was turned outside. She was watchful, always checking to see if another cow was nearby before she submitted to a friendly head scratch. During most of the year she was calm and placid, enjoying a chat with us or happily snoozing in the sunshine and chewing her cud. Calving time was another matter, and she turned from a happy cow into a harridan, fiercely and manically protective of her newborn calf.

I always found the business of calving rather nerve-wracking. In late summer and early autumn, we would check the herd twice or three times daily, watching them as they peacefully chewed their cud, looking for the signs that they were about to give birth.

A few days before a calf was born, we would see the cow's udder growing bigger, with the teats sticking straight down or out at an angle. Normally cattle have high hipbones ('hookbones'), but a few days before giving birth their pelvic ligaments start to soften so the top of the cow's tail sticks up a little more than usual.

The cow would isolate herself from the herd, finding a quiet spot to give birth. She would get up and down and start pawing the ground or stretching out her back legs as the contractions grew stronger. Two tiny hooves, often still in the bag of amniotic fluid, would appear and eventually with a great heave the calf would slide out onto the ground in a steaming rush of water. A healthy calf would take a first breath, and the mother would lick him clean and dry until he was strong enough to stand and take a first drink of colostrum milk from her udder.

Lop Ear always produced a healthy annual calf but became extremely protective of her baby, skittish and paranoid about anyone coming too close. She carefully hid her newborn in the thick layer of nettles at the side of the field, and the baby would sit with his legs tucked up beneath him, impervious to nettle stings as he lay protected from the weather and hidden from any predators.

Her calves were almost always the same colour, with a grey speckled hide shading to coal-black on the legs. We'd check on the herd each day, and from the safety of a quad bike watch the wobbly-legged tinies and their proud mothers snoozing on the turf. Lop Ear wouldn't let her calf come out of the nettles until we'd left, and the only way we could tell where she'd hidden her baby was to follow her by quad bike until she unwittingly revealed the location by rushing backwards and forwards in front of the nettle patch, mooing anxiously. I drove up and down in front to distract her and Steve would leap off the quad, gallop to the nettle patch, make a (very) quick check over of the tiny newborn and then leap back on the bike before Lop Ear made a run for him. We never allowed the general public into the field with calving cows as they were so unpredictable around people and dogs, and we couldn't guarantee anyone's safety.

Red Cow was usually one of the last to calve. Bizarrely, we always knew she'd started her labour by the way she stood by the water trough and pressed her forehead into the cool metal of the water tank. As her contractions became stronger, she would leave the trough and start pacing in a circle and stretching out her back legs one at a time.

One early autumn, Red Cow went into labour as normal and eventually gave birth in the far corner of the back field. At first light we noticed that there were *two* little bodies on the green turf. She'd had twins. They were both tiny, with beautiful, deep red coats and pale cream muzzles. One of the tiny twins was lying motionless and huddled in the thickest nettles by the wall as if Red Cow had been pushing him with her nose, trying to get him to stand up. She was now ignoring him, focusing all her attention on the other twin, which shook its head and mewled as she cleaned it with her tongue. I tiptoed over to the first calf and ran my hand down the little animal's wet flanks. I was absolutely shocked to see that it was dead, its body stone cold and rigid beneath my hands. This was the first time that I'd seen a dead baby animal, and I remember staring in horrified wonder at the long sweep of eyelashes fringing a deep black eye and the delicate shell-pink colour of the tongue. I couldn't understand why it wasn't alive. It looked absolutely perfect and was curled up as if it was just fast asleep on the bed of nettles. Fortunately, there wasn't time to think about it as Steve called me, and I had to turn away from the tiny body to help scoop up the other little calf twin who was bellowing with an open mouth as his mum fussed around him. Steve plopped him up onto the quad bike, and I sat on the back, clutching the wet body on my lap, facing Red Cow as she lumbered along behind, mooing to her little calf. We put the baby into some thick, dry straw in a stable and Red Cow followed him in, muttering and licking, trying her hardest to get him dry and make him stand up to drink.

Twins are not a welcome option. Sometimes they're too small, or too premature or the cow doesn't have enough milk. This little scrap seemed too tiny to survive but Steve sat in the straw, patiently batting away Red Cow's nose whilst he siphoned artificial colostrum through a stomach tube into the calf.

favourite game was to charge, helter-skelter, around the tree in the centre of the field, scampering alongside each other with their tufty tails sticking straight up in excitement. One would eventually stray a little too far towards me, and their mother would raise her head and watch in suspicion as the calf stretched out its tiny wet nose to sniff at my hand. I'd reach to stroke the calf's dished face and it would snort and jink away, thundering back to its mother and thrusting its head under her back legs, thumping her udder with its forehead until she let down her milk.

Calving didn't always go as planned. One year, the weather was wet and mild, and the grass grew lush and green and almost knee high. The cows put on weight at an incredible rate, and by the time autumn came around we knew that the calves they were carrying would be bigger than normal.

Henrietta was particularly worrying. She was a big creamy-coloured beast with a brown nose and legs and was carrying such a big calf she looked like a barrel. After a whole summer of over-eating her belly was tight as a drum, and it swayed from side to side as she waddled backwards and forwards to the water trough.

She went into labour one rainy September evening and began to push when the light started fading behind the trees at the bottom of the field. We watched her with binoculars from the back window, keeping an eye on her progress. After an hour she was still pushing, stretching her head and neck back and rolling her eye every time a contraction struck.

By this time Steve was poised at the back door, straining his eyes across the pasture, trying to see if there was any sign of the calf. He was always nervous around calving time, unable to sit down and relax, continually pacing backwards and forwards. So much could go wrong, and after nine months of carefully feeding and caring for the pregnant mothers-to-be we were on tenterhooks. Not only that, but each cow was valuable, a big investment, and each healthy calf was a chance to turn a much-needed profit.

After another half an hour, Steve couldn't cope any more.

'I'll have to go and see what's going on. Give me a hand with the quad and we'll take a look.'

By then it was pitch black and we needed the quad bike lights to see across the bumpy field. Henrietta was in the far corner, in a gap between an oak tree and the stone wall. She didn't get up when we puttered up to her on the bike, but simply looked around distractedly as we gently crept up to her side.

Steve quickly stripped off his top and poured gel all over his right hand. Reaching forward, he pushed his hand inside the cow, searching for the two tiny hooves that would show that the calf was in the right position. Henrietta groaned and started to push back against him, panting as sweat formed on her flanks. I was tasked with holding the torch and trying to keep it steady.

'What can you feel?' I asked.

Steve grunted and pushed forward a little further. 'The calf's all there but the head's back. It's stuck. And the shoulders are massive.'

The shoulders were too big to fit through the birth canal and the head had been pushed back, tipped so that the forehead was stuck against the pelvic rim.

Steve pulled out his hand with a squelch and, still naked from the waist up, went to find the calving jack on the quad.

A calving jack is a rather alarming-looking piece of equipment made from stainless steel and ropes that help to pull a stuck calf out of a labouring cow. The base of the jack is designed to fit over the cow's hip bones and ropes are attached to the calf's head and feet with a long lever that is cranked to literally jack out the obstinate newborn. It needs to be used carefully by someone who knows what they're doing.

Steve fed the rope inside the cow and pulled it gently around the back of the calf's head. Then he lubed up two more ropes and made two tiny loops which went above and below the calf's front fetlock (ankle) joints.

Henrietta was quiet, lying on her side with one ear cocked back and her eyes closed. Steve fitted the base of the jack against her backside and waited for a contraction to start building in her belly. We could see the muscles tightening against her skin, and as soon as the cow lifted her head in pain, Steve gently cranked the lever. The machine made a loud 'cronk, cronk' sound, just like a jack lifting a car when you need to change a wheel. Sweating and heaving, Steve pulled and cranked each time a contraction began and gently tugged on the head rope, trying to persuade the calf's forehead to drop down between the two legs. This was one of the first tricky births I'd seen, and I was torn between keeping the torch steady and wanting to sit at Henrietta's head, to soothe her as she grunted and groaned.

Two tiny, wet feet suddenly appeared at the entrance to the birth canal, tipped with a pair of black, rubbery hooves. Henrietta heaved herself up into a standing position, threatening the position of the jack. I had to put the torch between my teeth and help steady the base of the machine, as the lever tightened the ropes and the calf's head slowly started to slide out from under the tail.

With a great groan, Henrietta made a final heave, the shoulders popped out and the calf dropped unceremoniously to the grass, amniotic fluid spattering all over our legs and arms.

It was a little heifer, with a coal-black coat and a white splodge on her forehead.

Steve dropped the jack and rested his forehead against the side of the cow, his chest rising and falling as he tried to catch his breath. Henrietta stood with her head down, her sides covered with sweat; goop was dripping from beneath her tail onto the grass.

It was one of the most physically demanding things I'd seen Steve do on the farm. Calving jacks are a helpful bit of kit, but farmers still need brute strength to be able to drag a slippery calf through a too-small cervix.

'Christ, that was hard,' he gasped, as he wiped his nose against a shirt sleeve. Pulling on his jumper, he bent to inspect the calf, dragging his hands over its face to pull off any shreds of amniotic liquid. I helped him pick up the tiny heifer and carry it around to Henrietta's nose. She immediately bent her neck to sniff at the little animal and, pushing out her long pink tongue, started to lick the wet, black fur as the calf shook its head and flapped her ears at the unfamiliar feeling.

We gave Henrietta an antibiotic injection and a syringe full of painkillers and then stood back to watch them both. The black heifer was a big, strong calf that seemed none the worse for its traumatic introduction to the world. After a while she staggered to her feet, wobbly and shaky with stilt-like legs, and started to nose underneath Henrietta's belly for the first drink of warm milk.

We left Henrietta and her calf and drove back across the field, the cows watching us as we bumped across the curves of the pasture. I carried the calving jack into the house, as I'd need to wash and sterilise it before we could use it again, while Steve got straight into a hot shower, washing off the muck before collapsing into bed, exhausted.

Early the next morning he was up and out to check on Henrietta and her calf. She had tucked it into the bushes at the side of the field and she watched benignly as Steve carefully uncovered the animal and checked it was warm and had a full belly of milk.

Whilst we were involved in calving outside, we had grouped last year's bullocks into an unruly bachelor herd in the front field. They behaved like over-energetic teenagers, creeping up on anything interesting in their field, daring each other to get closer and closer, until, startled, one would spook and twist away, kicking with excitement as they all thundered into the distance.

Visitors to the farm used to find themselves in an unwitting game of Grandmother's Footsteps. A walker would start striding confidently across our field, often with a dog on a lead. The young bullocks would tiptoe along a few steps behind until the poor walker spotted that they were being followed by an alarming throng of snorting bullocks. We would watch from the window as the sweaty walker and wild-eyed dog made a panicky gallop for the stile and the young cattle careered in the opposite direction, mooing in excitement and pretending to be scared, kicking and bucking in hilarity as they scattered across the field.

We once watched an inspired rambler offer one persistent bullock a bite of his cheese sandwich. The animal couldn't believe his luck and gobbled down the whole slice, spitting out the cheese and almost swallowing the tin foil in excitement. After that, we'd often get a phone call from a perplexed visitor complaining that they couldn't get through the gate as there were fifteen black cows staring at him over the top rung, hoping for a spare sandwich. Our bullocks weren't dangerous, but just full of irrepressible curiosity.

One busy year, after she'd calved a little grey heifer, Lop Ear started to look unwell. She stood with her head down, well behind the rest of the herd, with her sides bloated like a balloon. She stamped every time her calf came near to drink and was obviously in pain, so we carefully herded them both out of the field and into a cattle pen near the cow shed.

Dave the Vet came to have a look. Dave is Scottish, very funny, and a brilliant large animal vet with one of the messiest vans I've ever seen in my life. He screeched into the yard and leapt down from his truck, already dressed in waterproof overalls. Opening the boot, he scrabbled around in the heap of medicine bottles, latex gloves and obscure lambing equipment until he found what he was looking for.

Lop Ear stood in the cattle pen, her head hanging low in misery and her tail firmly clamped against her quarters. Her tiny calf stuck close to her side, eyes wide in amazement as she stared at Dave and his medicine bag.

'Looks like bloat,' he said, as he felt along her side and pulled down her eyelids to look at the mucous membranes. Bloat is a really common form of indigestion that is caused by excess gas in the stomach and is often caused by eating too much rich grass in the spring or autumn months.

Steve held Lop Ear's head as I stood at the side, gently batting away the calf's nose when she tried to slip under the pen to get to the udder.

The vet quickly slipped a clear plastic tube down into Lop Ear's throat and poured in a thick 'anti-froth' liquid, which would hopefully dissipate the gas through the stomach tube. He put his ear against the tube top, listening carefully for the hiss of escaping gas.

We waited, but nothing happened.

'I'll have to go in through the side.'

Steve nodded. It would be more expensive but essential to quickly relieve the cow's swelling and pain. Steve held Lop Ear's head as Dave disinfected her flank and quickly injected a syringe full of anaesthetic into it. Grabbing a long needle and thin plastic tube, he poised his hand over her stomach, feeling for the best spot until he punched forwards, pushing the sharp tip directly into her stomach wall. Immediately there was a high-pitched whistle accompanied by a wretched smell as dark green liquid sprayed onto the cobbled floor. The fluid reeked of rotting cabbage and stomach acid, and I dropped my head to try to avoid the stench.

Coughing against the smell, I turned my head to see Steve's eyes roll back in their sockets as he slowly fell sideways with a heavy thump against the cobbles, directly underneath Lop Ear's head.

'Oh shit,' said Dave and let go of the tube, so that the fetid liquid sprayed all over my jacket.

I was now covered in a disgusting, green layer of foul vegetable matter, while Steve lay face down on the cobbles with Lop Ear sniffing interestedly at the back of his head, already feeling better from the release of pressure.

I rushed round, wiping smears off my glasses, and helped him up into a sitting position, his head dropping between his knees. He started to wake up, swearing lavishly and trying to push himself up from the cobbles.

Steve has a history of fainting when watching needles or even talking about operations or anything to do with blood. When I was pregnant, he lasted five minutes into our antenatal talk until the discussion of epidurals made him flee, retching, into the corridor. Talking to the nurses afterwards, they said that it was common for farmers to pass out when giving blood or having a jab. Steve is fine injecting an animal himself but standing by and watching someone else do the work makes him feel lightheaded.

'Aye, he always does that,' said Dave nonchalantly, ripping open a new needle and fitting it into the one of the biggest syringes I'd ever seen.

Steve groaned and started to get to his feet, swaying slightly, but as Dave started to inject a dose of antibiotics into Lop Ear, Steve noticed the enormous needle and suddenly sat down again. Eventually, the cow was well enough to be shooed into a warm straw pen with her calf, and Dave helped me to push Steve onto the quad so I could putter back to the house.

Lop Ear managed fine after her vet intervention and thankfully never came down with bloat again. Steve had a cup of sweet tea and a chocolate digestive and soon felt much better. Since then, whenever Dave visits the farm, Steve rushes to find a chair. Just in case.

Steve and I continued living happily together, and I grew more and more attached to the farm and its inhabitants. Cyril was enjoying his new life as a farm cat, spending hours perched on the very top of a tall stack of haybales, staring at the neighbourhood moggies or honing his hunting skills and proudly presenting a decapitated mouse or the gift of an unidentified rodent's intestines.

There were some aspects of rural living that I found difficult to understand. When I lived in the city, if something needed fixing I would use the Yellow Pages or Google to find a professional, ask them to do the work and then pay by cheque or bank transfer. Things didn't work like this on the farm.

If, for example, the boiler in the cottage wasn't working properly, I would ask Steve who to call.

'We've always used Norman the Plumber.'

'Norman the Plumber?' I'd ask in bewilderment.

'Yeah,' Steve replied. 'He's not actually a plumber as such, but everyone uses him and he's worked for us for years and years.'

'Is he expensive?'

'Nah,' said Steve and thought for a moment. 'Well, to be honest I've not seen one of his bills for a while so god knows. I think he sends invoices occasionally, when he can be arsed and his bank balance is looking a bit low.'

Norman's landline number was found, and he promised he'd be round 'when he'd got these few jobs out the way'. A week or so later he turned up on the farm with no warning and gave a perfunctory knock on our front door before immediately flinging it open and starting work on the boiler.

These working relationships seemed to have been going on for years and years: an almost feudal partnership between the farm and a circle of ageing odd-job men, who were trusted implicitly and occasionally turned up with an invoice or two, depending on the state of their personal finances.

One such man was called Ron the Mole. Ron was in his seventies, lived in the local village and had worked as a mole catcher with his dad and with his grandad. He didn't have a mobile or business card, and you could only catch him by ringing his long-suffering wife who would laboriously take down your plea for help and then pass it on to Ron. High House has always had a problem with moles: great beasties that produce massive heaps of spoil that appear literally overnight like huge, black daisy chains looping across our grass and crop fields. Mole hills aren't just unsightly; the exposed soil can infect grazing animals with clostridium, so each year we try to trap a few of the local population so they don't get out of hand. Ron was a leading light in the world of mole control and in great demand through the locality. He would turn up in his battered yellow van and park in the middle of the farmyard, clutching a handful of steel mole traps. Woe betide if he spotted you lurking by the hay bales as you'd then be stuck in a forty-minute conversation about neighbours and an in-depth discussion of lifestyles of farmers across the district. Finally he'd get to work, poking the ground for mole tunnels and pushing his steel rods into carefully chosen sections of turf. After a few days he'd come back and check his traps, eventually appearing at our front door with a sheaf of tiny black mole bodies, which he insisted on counting in front of us so we could see how many he'd caught and how much we owed him. Payment was worked out after a few minutes of careful sums on the back of a tobacco packet and was paid in cash. Ron was and is a true master of his craft, and everyone in the district is hoping desperately that he doesn't want to retire anytime soon.

Life for Steve and I settled into a comfortable rhythm as we worked alongside each other, out in the open air, in good weather and in bad, helping to manage the cattle and sheep. I learned how to pace myself through each day, choosing which jobs to do depending on the season and the urgency of the tasks. There were a lot of things I couldn't manage, such as driving the tractor (I'm too short to reach the brake pedal, and we had to buy an automatic quad as my tiny feet couldn't manage the pedal clutch) but I plugged away, learning what I could and becoming an integral partner in the day-to-day work of the farm. Everything ran smoothly, and without any hitches or catastrophes until 2001, when our happy, peaceful life skidded to a stop with the eruption of foot and mouth disease at a small farm down the road, at Heddon-on-the-Wall in Northumberland.

CHAPTER THREE

Foot and Mouth

*The farmer has to be an optimist or
he wouldn't still be a farmer.*

Will Rogers

The last epidemic of foot and mouth disease in the UK happened in 2001. Writing about this awful episode brings back the deep feelings of bitterness and overwhelming hopelessness that we felt during those ten long months. In a way, the isolation and loneliness of that year were replicated during the Covid crisis, and I can see how the tragedy of both events will throw long-lasting tendrils of trauma into everyday lives.

When the news of the outbreak was first reported, Steve and I were sitting watching the evening news.

'Bloody hell,' he said, reaching for the volume control on the little television in the kitchen, 'listen to this!'

The reporter announced that foot and mouth had been found at a pig farm, just down the road from us in the village of Heddon-on-the-Wall.

'I remember Dad talking about the outbreak in the 1960s,' said Steve, staring at the figures on the screen: men in white biohazard suits spraying disinfectant on the walls of a shed and driving tractors with front loaders full of dead pigs. 'They have to stop it spreading.'

In the following days it became clear they wouldn't. This outbreak was on another level altogether.

The disease spread fast. The first signs of an infected animal were red-raw, open blisters on their mouth or tongue, a drooling mouth, or a limping sheep or cow, bleeding from sores between the cleats on their hooves. Every morning, whilst doing the rounds, Steve or his dad would anxiously check for any limping animals or sheep that wouldn't eat their feed. In a way, the absolute uncertainty caused the most emotional distress and trauma. Every day, whilst walking the fields or feeding the cows, they were on high alert to anything that suggested infection with the disease, such as a lame animal or a cow with a frothing mouth.

One day, Steve burst through the front door and shouted for a hand. Pulling on my coat, I saw that he had a sheep in the little trailer on the back of the quad bike. She was lying down on the bottom of the trailer and her head hung in the straw.

'Give us a hand whilst I check her feet,' said Steve, and I quickly stepped in next to the ewe and helped him roll her onto her side. My heart started pounding as I saw that she was frothing at the mouth.

Steve carefully checked each foot and pulled out some of the mud and grass lodged in between her hooves. She winced and tried to pull her leg back, but he gently clipped the foot wall and sprayed some purple antiseptic into the middle.

'No blisters,' he said shortly, and my breathing slowed down a little.

We carefully cleared the froth from her mouth and opened her jaw. Steve checked for red marks or sore patches across her tongue and the roof of her mouth. There was nothing there. Relieved, we hauled her into the shed and made her comfortable in a pen.

'I'll have to call the vet,' said Steve, chewing his lip. In silence, we walked back to the house, and he rang the Ministry Veterinary Department.

'I don't think it's foot and mouth,' Steve said, tensely. 'But can you come out and have a look?'

When the Ministry van pulled into our farmyard, we saw that Dave, our own vet, had arrived as well. They both climbed into biohazard suits, dipped their welly boots into the bucket of disinfectant and scrubbed their hands with antibacterial soap.

There was no shaking of hands, and we quickly trooped into the shed whilst the vet took his torch and carefully inspected the ewe. I had balled up my hands in my pockets, fingernails pressing against my palms as we waited in almost unbearable anxiety for the prognosis.

After a while he straightened up. 'Nope, you're good. No foot and mouth.'

Dave let out a huge sigh, and Steve gave the thumbs up, although I noticed his hands were shaky. The ewe had a case of pneumonia and only needed a few days in the warm and an injection of antibiotics.

My stress levels started to subside as both vets, smiling now, came out of the shed and started the long process of disinfection and decontamination of their boots, hands and vehicles.

'Thank Christ for that, eh?' said Dave, as he poured disinfectant over the wheels of the van. Steve, still speechless with anxiety, nodded his agreement and helped him spray down the wheels with hot water. That night we both had a glass of whisky to calm down from the shock of it all. Thankfully, that would be the only time we had to call a Ministry vet throughout the whole outbreak.

For those poor souls who did become infected, the process was swift and brutal. Once a Ministry of Agriculture vet had confirmed an infection the culling began, and a trail of vets and agency workers swung onto the farm with the aim of killing every single sheep, cow and pig on the place. All the animals that could possibly carry or pass on the disease were slaughtered, immediately.

We sat transfixed by the evening news as they filmed the piles of burning carcasses and the obscene pyres that were built all over Northumberland and Cumbria to cremate the vast number of dead animals. Vets had to slaughter hundreds upon hundreds of fit and healthy in-lamb ewes, and we heard of whole flocks of sheep with newborns being systematically culled. Entire pedigree flocks of sheep and herds of cattle were killed. Newborn calves were shot.

The ordeal of it all caused mental breakdowns, suicides and huge rifts within families. It caused great collective trauma and suffering not only to the farmers, but also to the vets and agency workers tasked to carry out the culls.

We had the radio on constantly, listening to the updates on the spread and how the infection was creeping closer and closer to our farm. It felt as if we were standing on a precipice, waiting for the day when we would spot a cow off her feed, or a calf struggling to suck, and then we'd quickly become one of the fallen, the next farm to face this holocaust.

I distinctly remember driving home along the A69 and seeing huge mounds of dead cattle in the field adjoining the road. You couldn't make out the individual animals amongst the morass of blackened carcasses, but you'd spot the occasional stiff, charred leg that poked out from the massive pyres of dead animals. The smell was the worst: as the burning smoke drifted across the countryside, a choking bitter barbecue stench permeated everyone's clothes and hair.

After a few months, Steve was at his wits' end. We were not allowed, by law, to move sheep and cows from field to field. We couldn't sell any lambs or young cattle at the mart as it was shut down by the government. Grass became scarce due to the sheer number of animals that we needed to feed. All the nearby farmers were in the same boat, and we all began to run up massive bills as we ran out of animal fodder and had to buy in hay and straw to feed and bed up the animals.

Worse was to come. Steve took a phone call and learned that Simon (note: not his real name), a farmer from a nearby village, had confirmation that his herd was infected. Simon had spent his entire life on the farm, first as a young boy helping his father and then as sole proprietor, building up and caring for a pedigree herd of beef cattle and a small flock of Cheviot sheep. He'd never married and lived alone. The fact that he was the first in our neighbourhood to suffer an outbreak absolutely flattened the poor man. Tears running down his face, and by turns angry and despairing, he watched as the vets and soldiers drove onto his farm the very day that the disease was found in his cattle. He stood outside the big hemmel and watched as his entire herd of prized suckler cows was shot. The worst horror was that he was still in the middle of lambing. He was given the option of staying in the house, but instead, he helped shoo all his much-loved pregnant ewes and all the new mothers and still-wet newborn lambs out of the warm shed and into the rain and wind where they were pushed through a sheep race and were shot, one by one, through the forehead by a vet wielding a bolt gun. We could hear the panicked bleats of the ewes and their lambs drifting across the valley. I still cry when I think of that poor, brave man trying his very best to look after his animals right to the very end.

Simon never went back to farming, and he never really fully recovered. He put his farm up for sale and moved away to live with his sister. The buildings and land were snapped up by a developer and turned into executive barn conversions.

We escaped by a whisker. Even though the infected farm was nearby, the government hadn't yet decided to mandate contiguous culling, a policy whereby not only the animals on infected farms, but also *all* animals on adjacent farms, were slaughtered, regardless of whether infection had been reported there.

Steve became very withdrawn. In those days there was no social media, no chance to share problems online. He couldn't go to see friends, or nip to the pub for a quick pint and have a chat or share his worries with the rest of the community. If he had to leave the farm, there was the long, laborious washing and disinfecting of his car and trailer, boots and clothes. So, he rarely bothered, growing an epic beard and living off deliveries from the local shop. This went on for months and months and months. Today, we can see the echoes of that time during the Covid crisis, but it's different. In those days the majority of people in the UK went about their life with no real change or understanding of what was unfolding in the countryside. It felt doubly isolating, not only to be kept apart from friends and family but also to see the distance that separated us from those who didn't comprehend what was happening to the farms in their country.

But life went on. Spring turned to summer then merged into autumn, and it was time to put the cows to the bull. Our fields were still full of last year's unsold and unwanted lambs and bullocks, but the farming year can't wait, and we knew that our cows needed to be in calf for next year. Bully was no longer able to work due to his age, and Fred wasn't physically able to serve all the cows, so Steve had to reluctantly ring around until he was able to find a farmer who was happy to rent him the use of a suitable bull.

The animal was an hour up the A1 on the outskirts of Berwick, but foot and mouth still raged in the region, and we realised it might be impossible to move a new bull onto the farm. However, after a frantic evening of ringing Trading Standards and government vets, we had a solution.

One bright Sunday, Steve began by sterilising and power washing every inch of his trusty old trailer. A rather nervous Ministry vet appeared on the farm, anxiously checking his paperwork as we started the laborious process of disinfecting our boots and pick-up truck.

Two serious-looking men from Trading Standards arrived in a white van and stood to one side as we finished sloshing disinfectant around the farmyard. Eventually, our dismal little cavalcade started up: Steve in front with a rattling trailer, then the grave-looking officers from Trading Standards and finally the worried-looking vet bringing up the rear.

We pulled up at a lay-by just outside Berwick and saw an apprehensive farmer pacing up and down by a spotless cattle truck. We pulled in and went through the entire laborious process once again, washing and disinfecting and sprinkling sterile sawdust to cover the floor of our trailer.

The new bull, a massive beast with a completely black hide, was loaded into our trailer and Steve exchanged a few terse words with the Scottish farmer. They didn't even shake hands, to prevent any possibility of infection, something we all recognise so well now.

The bull stood quietly in the back, peering out of the slats of the trailer as we pulled away, back to Northumberland, the Trading Standards guys and vets puttering along behind us. We could only drive around fifty miles an hour and many cars whizzed past, passengers staring in interest at the heavily loaded trailer and the bull squinting out of the side.

Our sad little convoy hit the outskirts of Morpeth, and Steve suddenly noticed the Trading Standards van frantically flashing its lights. It pulled up beside him in the next lane, with the officer gesticulating wildly for him to move onto the hard shoulder and stop. Steve screeched to a halt and wound down the window. The officer ran up, breathless and agitated, yelling, 'You'll have to stop! The bull's been pissing on the road!'

We looked behind us and spotted the incriminating, long widdle of urine that trailed down the slow lane of the A1. The new bull had obviously done such a copious wee that it had overwhelmed the sawdust barrier and dribbled out of the trailer onto the tarmac.

I goggled at the Trading Standards officer, as Steve rubbed his hand through his hair.

'What exactly do you want me to do about it?' he said.

At this point, the stress of the journey and the ridiculousness of the situation hit us both, and we started to giggle, snorting a little wildly through our noses as the official hopped from one foot to another.

'You'll have to add more sawdust to soak it up,' he said.

Steve was quiet for a second.

'You want me to get *into* the trailer *with that bull* and throw sawdust around?' he asked in disbelief.

The Trading Standards officer nodded, his chin jutting as he avoided Steve's eyes.

Steve went into the back of his pick-up and pulled out another bale of sawdust, clambered over the breast bar into the trailer next to a rather surprised bull and scattered the sawdust onto the floor. Luckily the bull seemed very quiet and just moved to sniff the newly clean floor.

When he emerged from the trailer the officer was holding a green watering can. I watched in amazement as he thrust it at Steve.

'Go up the road and wash it down with this,' he barked. He pointed inside the can. 'It's disinfectant.'

'What? I'll be flattened!' snapped Steve as he turned to watch cars whizzing inches past the trailer, making the pick-up truck and heavily laden trailer rock violently.

'If you don't, we'll stop you moving the bull.'

So, red-faced and furious, Steve walked back up the hard shoulder, dodging cars and trucks and throwing pink frothy disinfectant over the long line of wee.

He stalked back to his car, stiff-faced and fuming, stepped in and slammed the door hard. I slid back into the passenger seat, and then, without waiting to see if Trading Standards and the vet were close behind, Steve pulled out into the stream of traffic. The journey carried on in silence until we all arrived at the farm.

After washing the wheels and our boots again, the placid, new bull was walked down the ramp into a clean cattle pen, and Steve marched round to the power washer to clean the dirty sawdust and bull muck straight into the farm drain.

The Trading Standards officers checked off a few boxes on a form and drove off. The poor vet hung around a little longer and once the power washing had finished, waited another moment and apologised. Tempers had run extremely high.

On a positive note, our new bull remained disease free and did a great job, enthusiastically joining Fred in mating with our cows and siring some healthy calves.

But still the foot and mouth drama dragged on. Eventually local farmers couldn't feed their animals and became so desperate to move their stock they asked the government to slaughter 700,000 of them, as they were stranded in fields or on farms in increasingly overcrowded conditions and could no longer be properly cared for. The government was overwhelmed and eventually cooked up a new system called the 'Slaughter Scheme', whereby farmers could sell their lambs, pigs or calves to the government for a set amount. Steve sent almost all his lambs to this scheme, desperate to avoid even bigger food bills, but only received a small proportion of their actual worth.

After a few more despairing months the foot and mouth crisis was finally contained in a few isolated areas and we emerged, blinking into the light, to discover that High House Farm had survived.

But the local community was in turmoil, with rumours of farmers and dealers deliberately infecting their stock so they could be 'killed out' and therefore claim insurance from the government. Poisonous stories proliferated, shared from one farm to another or whispered in the local pub, with filthy looks and the oft-repeated line, 'Oh, it's alright for *them*, *they* claimed the compensation, *they'll* be rolling in it.'

To be fair, the insurance claims system *was* flawed, allowing farmers to appoint their own valuer to assess their culled animals rather than using an independent, impartial company. The thinking was that this led to corruption, where insurers valued the culled animals above their actual value, marking them as pedigree when the animals were just straight forward commercial 'fat' stock.

In addition, some of the farmers who had their flocks and herds decimated by foot and mouth were then employed by the government to go and clean other infected farms.

The farming community is full of rumour and gossip at the best of times, but the foot and mouth outbreak set it on fire, and jealousy and anxiety fuelled gossip about dishonest or dangerous practices.

When it was over, the farmers whose farms had been infected were able to claim compensation and use the money to buy new stock, whilst those who had escaped the disease were left with a glut of animals and huge feed and bedding bills that crippled many of them and caused a number of them to go under.

Steve was left out of pocket by nearly £100,000 and it took almost eleven years to pay the bills that resulted from the crisis. We were doubly affected as our stock weren't pedigree breeding animals but rather run-of-the-mill 'meat' stock, and literally nobody wanted to buy them as the market was saturated with the backlog of nine months of unsold animals. The glut of animals and the huge feed and bedding bills nearly dragged us under.

Our farm and the local community are still recovering from the epidemic. It's not an exaggeration to say that many still suffer PTSD from that time. Visions of piles of dead and dying sheep or enormous Dante-esque pyres of burnt cattle still intrude into normal life. It left behind poisonous rivulets of jealousy, bitter speculation about corruption within the industry and frustration with the ill-thought-out policies that were implemented to contain the epidemic.

As for Steve, at one low point he admitted that he almost wished the farm *had* got the disease. There was no compensation for the huge bills that we had racked up. His animals were sold at a fraction of their price and the fields of grass were stripped bare.

He told me one day at the dinner table that he had decided that 2002 would be his final year of keeping beef cattle. He propped his head in his hands, staring at a long list of accounts, in shock at the number of invoices to be settled.

'We just can't afford to keep them. We can't make any headway against these bills.'

His finger stopped at a particularly gigantic feed bill from a local merchant.

Standing up, he looked out of the window at the herd of cattle standing around the feeder, peacefully chewing on a newly opened bale of silage.

'We'll have to sell them,' he said hopelessly. 'I'll stick to sheep. At least it won't cost a fortune to feed them over the winter.' He was trying very hard not to cry and rubbed his hands over his eyes, blinking back the tears as he stared out of the window. Shrugging on a coat, he went outside to feed the sheep and didn't come back into the house for the rest of the day. Once Steve makes a decision, he very rarely changes his mind, and he was determined that we would sell the herd and stop the continual drain of money from our depleted bank account.

And that was how the last cattle herd disappeared from High House Farm. It wasn't an overnight process of course. We still had one more calving to do, but in September 2003 all our cows (with calves at foot) were sold to other farmers at the mart. Steve stayed in the auction ring with every single beast, patting their rumps as they trotted obediently out of the ring. The money raised from the sale went towards the outstanding bills. Not a penny was left over.

After the cattle left, all the sheds and fields were turned over to sheep and we prayed that there would never be another foot and mouth outbreak in Northumberland. It was now time to look to the future, to try to put the trauma behind us as we started on the next chapter for our farm.

CHAPTER FOUR

Beer and Brides

Steve and I were married on a balmy autumn day at an ancient, grey stone church in the village of Corbridge. We were both so excited and nervous and when the actual day dawned, and it was time to climb into my dress, I still couldn't believe it was actually happening. My wedding gown was made to measure and involved a perfect cloud of ivory lace over a very tight satin corset, which made it almost impossible for me to do anything apart from lean picturesquely against a chair. My bridesmaids were dressed in gold satin and carried gigantic bouquets of chrysanthemums in hues of bronze and crimson, which were carefully chosen to complement the theme of an autumn wedding.

The wedding ceremony was booked to start at two o'clock, and ten minutes after two I reached the church and passed under the medieval carved archway, nervously clutching my dad's arm, to the rousing sounds of a trumpet voluntary. Halfway down the aisle I suddenly realised my dad's cravat wasn't properly tied and that I was actually, seriously getting married to my best friend. And, as usual in times of high emotion, I promptly burst into tears. The rest of the service was punctuated by my occasional happy sniffles and my dad and brother booming out the hymns from the front pew. It was a wonderful, jaw-dropping day. It involved a full choir, a real-life castle, vintage Rolls-Royces, a wedding cake with a tiny model of a sheepdog, Cyril the cat and millions of photographs. It seemed to pass by in a flash, but it was an authentic, rousing Northumbrian wedding, filled with love and speeches, real ale, dancing and far too many profiteroles. We had to get straight back to work on the farm the very next day, but it was a wonderful start to a very happy marriage. I think experiencing my own wedding day, and the full complement of emotions from nerves to tears and laughter, made it easier for me to see how all the little details pull together to make a faultless whole. This stood me in great stead when we were trying to decide what to do with the future of the farm.

We'd been desperately casting around for ideas for new businesses, but after the foot and mouth crisis diversification became imperative. Our financial situation was becoming perilous, and we needed a new money-making idea to bolster our ever-reducing cash flow.

In 2002, the government implemented the Rural Development Fund which was set up to help the countryside recover from the crisis. It was designed to encourage farmers to think outside the box and come up with new ideas for making money from their land and buildings. For smaller family farms, the parlous state of the traditional farming industry meant that they had to find a way to diversify, or they wouldn't survive. Margins were tighter, costs were rising, and lamb and beef prices were dropping.

Steve had always been interested in real ale and had made a few attempts at home brewing: the type of kit that involved big glass jars and rubber tubes and sat in our airing cupboard belching bubbles of carbon dioxide. After doing lots of reading around the subject, he came up with the idea of setting up a real ale microbrewery on our farm and using a percentage of our barley crops as the malt ingredient. I was less enthusiastic as my palate is firmly sugar based, and if a drink isn't brightly coloured and garnished with a bit of fruit and a straw, I'm not keen. (In fact, I still hate anything with a 'grown up' or bitter flavour. Coffee, beer, lager and red wine all taste a little bit like burnt water to my palate, which obviously hasn't matured since 1992, when I first discovered a sickly alcopop drink.)

A short evening course in microbrewing only fanned the flames of Steve's real ale obsession. He gathered some plans together and started looking at a suitable location on the farm for a brewing plant. As it happened, the original farm buildings were ripe for renovation. The farmstead was built in 1840 and many of the main buildings had fallen into disuse as the old cart sheds and their entrance archways weren't big enough for modern tractors. Some of the old stables were still used during lambing or to house cows and calves, but the old threshing floor and grain stores stood teetering into dereliction, filled only with old, rusty implements and sacks of dusty grain.

One evening during the summer of 2003 Steve and I stood at the top of the rickety stairs in the old threshing shed and stared at a massive hole that dropped straight through the roof tiles, then through rotten floorboards, onto the cobbles in the stable below. Everything was covered with a thick layer of pigeon muck mixed in with feathers and pieces of old plaster from the ceiling above. We skirted the worst of the rotten floors and tentatively tiptoed across the floor. I drew my hand across the cobwebbed walls, clearing enough of the dirt and dust so I could see the stone below.

'It's a lovely colour,' I said, peering closely at the yellow sandstone, still bearing the chisel marks of the nineteenth-century stonemasons, 'and there's a fair bit of space up here.'

Steve kicked at a pile of empty feed sacks. The dust rose in huge, choking clouds, and he started to cough.

'It needs a hell of a lot of work,' he said, 'those floors and window frames are rotten.' Some hung at drunken angles, their glass long shattered on the cobbled yard below.

I started poking around, an idea forming on how we could use the buildings to complement the brewery and bring in a little income on the side. 'Why don't we turn this part of the buildings into a tearoom and restaurant, and you can run the brewery alongside?'

Steve looked unconvinced.

'It'd bring more people to the place, and if they come for a cup of tea they're more likely to buy some beer to take home with them,' I said, warming further to my new idea as I poked around in the corners of the room and paced out the floorboards to measure the space. 'We could have weddings and receptions and parties and things.'

I started sifting through the piles of old machinery and emerged with an ancient scythe and a rusty horse bit, still with the reins attached.

'Least we've got something to decorate the place with,' I said, waving the wobbly, rusty blade.

'Aye, there's all sorts in here. We need to get rid of all the pigeon crap first though.'

It was almost impossible to see how the finished building would look, but after taking careful measurements we drew up some plans for the layout of a restaurant, a ten-barrel brewing plant, a tiny malt loft and a titchy kitchen.

Opposite our old farm buildings stood an open-sided haybarn, built in glorious, creamy sandstone with cast-iron columns and a Welsh slate roof. It was built in the mid-nineteenth century and we still used the building to store the hay and straw over the winter. It was a beautiful example of a Victorian 'Dutch barn' and had some very picturesque ventilation openings on the rear wall that looked just like medieval arrow slits. As it was Grade II listed we couldn't alter any of the original building, but we decided that with the addition of a decked floor, some stairs and electricity it would be the perfect backdrop to hold romantic outside wedding ceremonies.

The banks and business advisors fell over themselves to help. In those days they were enormously enthusiastic about turning derelict farm buildings into a working business, and the buzzwords of 'rural employment' and 'repurposed agricultural sheds' rang across many a meeting table.

We were given a European Rural Development Programme grant for 40 per cent of the initial building work, and the rest came from loans and a fairly substantial mortgage. I was tasked with writing a business plan for the tearoom and shop, and I spent hours looking up average tourist footfall and income per visitor. I discovered that providing teas and coffees for passing trade isn't massively profitable, and if we wanted to make this new venture a success we'd have to rely on attracting groups and special-interest visits to boost our income.

Steve was working hard on the brewery, and he'd taken to scouring the internet looking for secondhand equipment. Due to the peculiar shape of the grain store, and the fact that we had to add insulation and line the walls with sheets of food-grade plastic, each piece of brewing equipment had to be made to measure to fit in the ever-decreasing space. Costs were mounting, and I looked nervously at our bank statement each month, trying to foresee a time when we wouldn't be permanently in the red.

Gradually, the farm buildings were restored to their old glory. All the stonework was repointed, the floorboards were repaired and stripped back to the original wood, and the huge oak beams were cleaned of layers of old whitewash and treated for woodworm.

The golden sandstone looked breathtaking against the grey wood beams and the pale-coloured floors. There were lots of original features, including a recent find: in one corner of the new tearoom we discovered a set of pigeonholes hidden behind a thick layer of cardboard. Pulling the old damp cover away from the wall we discovered eight tiny little cubby holes, which would have housed dove or pigeon nests, the eggs and squabs being an important source of protein in those Victorian days. They made brilliant shelves and I made a mental note to include them in my plan for a miniature shop so we could display the different bottled beers that we were going to have available.

We bought a secondhand bar and installed handpumps and drinks fridges. After many trips to Ikea, we had amassed enough tables and chairs for the restaurant, and we turned the downstairs calf pens into a newly fitted commercial kitchen with an impressively fast dishwasher. The toilets were built into a blocked-off, curved doorway and the bottling plant and a bonded warehouse were created from the haystore downstairs. It was incredible to see what the builders could create from our dark, straw-filled calf pens and filthy, cobwebbed stables. The local craftsmen were wonderful to work with. They were patient to a fault, immensely skilled and extremely hardworking. One happy day, when the internal stone pointing was almost finished, we found all our masons clustered around the bottom of a ladder, propped against the highest beam in our newly repointed malt loft.

'What you up to?' I said, tipping my head back to see Jim, the chief stonemason, teetering at the top of the ladder.

'We're bringing you good luck,' said one of the lads, grinning out of a face turned white with plaster dust.

Jim was now using his finger to press something carefully into the wet plaster. He turned and came back down the ladder.

'Go up and have a look!' he said, wiping his hands against his dusty overalls.

I made a face at him, and he smiled back blankly, refusing to tell me what he'd done. So I climbed slowly up, peering at the wall and stopping when I reached the bottom of the first beam.

In the wall next to the hole where the wooden beam slotted into the wall, Jim had pushed two copper pennies into the damp plaster and had used his fingernail to mark the date and our initials into the paint below.

'It's for good luck, see?' he shouted from the floor below. 'Helps keep all the bad spirits and witches away.'

I laughed, delighted at the idea, and traced my finger around our initials. There had always been a tradition in this region of using hastily scratched 'witch marks', candle burn marks or bits of iron to protect new farm buildings against common disasters like fires or floods. I loved the fact that we were using ancient methods to defend our new brewery against catastrophes.

The pennies and the initials are still there, and on sunny days you can see them high above you, winking in the light and continuing to guard against bad days and evil thoughts.

Steve had already started brewing his very first casks of ale and spent hours tinkering with recipes in the fermenting room, with handfuls of test tubes and thermometers. He turned out to be a careful and consistent brewer, and the new beers gained a reputation for settling well and being full of flavour. I spent many days on the phone ringing landlords and pub owners to persuade them to try a cask, and it became easier when customer feedback on our ales trickled through, especially as it was overwhelmingly positive.

With help from a local graphic designer, we designed a new logo – with two sheafs of hop plants curling around the words 'High House Farm' – and created a basic website with our opening hours and new tearoom menu. We were ready to go.

I'm aware that I've made the process of creating a new business sound really smooth and uneventful, but in reality there is so much work behind the scenes that if I'd known the extent of the rules and regulations that we had to research and follow, I might have suggested that we diversified into an industry which was less tied up in red tape. Building nuclear power stations, perhaps? As well as Customs and Excise I had visits from Trading Standards, Health and Safety, COSHH Duty Inspectors, planning departments, the Food Standards Agency and many more. Each had complicated regulations that I had to demonstrate I was following before the brewery, wedding venue and tearoom were allowed to open.

Eventually, we reached the end of the red tape and rather gingerly opened to the public.

It was all a bit of a culture shock. Steve had never dealt directly with customers and spent a lot of time hiding in the office in case someone asked him something, and whilst I'd worked in shops and as a bartender in the past, managing an entire wedding venue, tearoom and brewery on top of being constantly accessible to the general public was very different. I had hoped that Steve and I would be able to cope with the tearoom on our own, but it was soon apparent that we needed more staff. Persuading someone to work for us was another nightmare as we're not on any bus routes, all staff had to be over eighteen years of age to serve alcohol and needed to be free to work every weekend and through the holidays.

Fortunately, I found a lovely lady called Nina. She was nineteen years old, from Ukraine and had moved to Northumberland to move in with her boyfriend. She lived in Hexham, had long blonde hair and even longer legs and spent a lot of time telling me how disappointing she found the Northumbrian scenery, weather, nightlife and shopping. She hardly ever mentioned her boyfriend at all, apart from showing us the new jewellery he'd bought her after their latest shopping spree.

Nina made a great addition to the brewery, especially serving beer to the local farming contractors who came for a pint and to stare longingly at her across the bar. She tossed her blonde hair flirtatiously, batted her long eyelashes and spent ages teaching them some absolutely filthy Ukrainian swear words. (I really hope that in tractors across Northumberland the word 'korva!' is still ringing out when the drivers realise that their drill has bounced off a large stone.) Nina was a breath of fresh air. She was patient, hardworking and had a raucous sense of humour. She drove like she was being chased by a herd of wildebeest and you could always tell she was in the car park due to the sound of tyres screeching as she barrelled her car round the corner. If she found something irritating there would be a long, muttered monologue behind the bar, peppered with mangled English swear words and lengthy Ukrainian curses.

The fact that our new tearoom and wedding venue were slap bang in the middle of the farm meant that we weren't always able to hide the more unsavoury parts of country living. The local knacker man always seemed to time his visits to our farm on very warm days to pick up any fallen stock, usually right in the middle of a wedding ceremony, and he seemed to take great delight in revving into the car park with his smelly deceased cargo, leaving a rainbow of diesel fumes and pungent gusts of decomposing animal carcasses.

Our new tearoom looked over the back of our sheep sheds, so that visitors could peer out and hopefully see an idyllic view of farming – all cuddly lambs and peaceful snoozing ewes. I hadn't thought about the fact that it also meant that our customers would have a ringside seat to any day-to-day catastrophes.

One morning at feeding time, an elderly ewe who had been calmly chewing her cud the evening before abruptly keeled over in the straw just underneath the tearoom window. She lay on her side, legs stretched out stiff, whilst Steve and I stood in the straw beside her and debated what to do with her carcass.

'I haven't got time to move her, and the knacker man won't come 'til after lunch,' said Steve, scratching his head with one hand. 'I could cover her up with a tarp, but it'll be really obvious – everyone will know there's a dead body under the sheet.'

I looked up at the tearoom window, the glass glinting in the light of the sunrise, trying to gauge how much people would be able to see from their seats inside.

I patted the dead ewe with my hand. 'She looks as if she's asleep. Sort of,' I said. 'Maybe if we just leave her people will think she's having a nap?'

Steve looked unconvinced, but with no better idea, I adjusted the ewe's head so it wasn't at such an awkward angle and went to wash my hands and change into my waitress uniform.

During the morning, I kept glancing out of the window. From a distance, with half-closed eyes, the sheep did look like she was fast asleep with her legs flung out in sleepy abandon. As the morning crept on, the ewe's body began to bloat. Animal carcasses do this very quickly, especially in warm weather, and I began to curse my idea of just leaving her and hoping no one would notice.

By lunchtime the poor ewe was completely spherical with gas and had rolled over onto her back so that her legs stuck straight into the air. She was like a round ball with four little leggy sticks at each corner. No one could possibly think she was asleep. All the other ewes carried on around her, unconcernedly munching on hay or chewing their cud, pointedly ignoring the deceased member of their little flock.

'I don't think that sheep's very well,' pointed out one elderly visitor as she peered out over the straw pen.

'No. I think it's feeling a bit poorly,' I agreed, as I tried to inch in front of the window to block the view.

'What's the matter with it?' the old lady said, trying to lean around me to get a better look.

I managed to distract the curious lady with a fresh pot of tea and a look at the dessert menu. By the time she'd chosen a slice of chocolate cake the knacker man had thumped down the back of his lorry, attached a large hook to the dead ewe and pulled her out of the shed, winching her up the lorry ramp with a squeal of metal rope. It took about thirty seconds. He shoved the paperwork into a hole in the side of the shed, jumped back into the cab and roared down the farm drive and back out to his next pickup. The sheep joined a heap of other dead animals – cows and calves and lambs – all collected from farms right round Northumberland. (Note: It sounds horrible but it's a very slick and professional system, with dead animals collected twenty-four hours before they can spread disease to the rest of their herd or flock. The carcasses are taken away to be ground down into fertiliser or burnt in a biomass power station, to ensure that nothing re-enters the animal or human food chain.)

As the summer rolled on, I began to wish that we'd set up our tearoom as far away from the farm steading as possible. One very warm and sunny day, I was clearing teacups off tables and staring vaguely over our sheep pens. Looking out of the window, I noticed a cute little mouse sitting on his haunches on one of the piles of wood in the barn outside. It was busily cleaning its tiny whiskers with a pair of sweet little front paws. I've always loved field mice and don't mind seeing them around the buildings at all, so when Steve wandered past, I pointed out the little fluffy creature.

He leant across me and raised his fist to bang loudly on the window glass. The sudden noise made the animal scamper down the pile and bolt underneath the wood, and when it moved I caught a glimpse of a long, hairless, creamy pink tail.

'That's a great big rat,' he said, rubbing the side of his fist.

'It was pretty cute,' I replied, oblivious to Steve's worried frown.

I completely forgot about the rat appearance until I was cashing up the next evening. I glanced out of the window above the bar and spotted a flurry of movement in the straw below. As well as a couple of bigger brown rats sitting on the pile of wood, there was also one investigating a bucket and another two or three on a bale of straw, leaping in and out of holes in the side.

I called Nina to come and look, and she shuddered.

'Oh, bloody hells, rats,' she said, with her blue eyes scrunched up in disgust.

'There's so many of them,' I said in wonder. I still thought they looked sweet, with their playful jumps and bounces and the way they sat to clean their noses and whiskers.

'Where is the poison?' asked Nina. 'You need to kill them all. They grow and grow…' She trailed off, her mouth turned down with revulsion.

I decided to go and have a look myself, so later on, after closing up, I walked around to the shed and flicked on the lights as I unbolted the door.

I screeched as the floor was suddenly alive with a thick carpet of brown rats. At the first glow of a light, they swarmed and ran in their hundreds, jinking over the straw, across the bales of hay and bags of feed and down into the sides of the building. I remembered seeing a TV documentary about a plague of rats in Australia, and this looked exactly the same: the writhing heap of rodents, the way they scuttled across the floor and jumped frantically in a panic to escape.

I turned around and ran, leaving the door to bang behind me as I went to find Steve.

We already paid Colin, our local pest control man, to appear every few months to replenish the black plastic boxes full of rat poison and to check for signs of infestation. But he'd only visited a couple of months ago, so surely there couldn't be an actual rat invasion in such a short space of time?

A more pressing worry was the fact that we were scheduled to hold a wedding reception the next day, and I had visions of guests screaming as they came face to face with some enormous, yellow-toothed rodent interloper. The tearoom was in a completely different building to the shed, but what if they tunnelled into the drains? Or made a ratty expedition up the waste pipes into the loos?

Colin had promised to come out the next day, so all we could do was turn off the lights and make sure all the doors were closed and the feed bins were locked. I lay in bed that night with my feet scrunched up towards my middle, my brain full of thoughts of ratty tsunamis.

I didn't dare check the shed the next day but concentrated on making sure the wedding decorations were perfect and the table plan was all properly set out. As I folded napkins, I spotted a few brown shapes slinking around the floor of the shed below and when the music was turned down, I swore I could hear little scrapping claw noises under the floor.

'This is no good,' announced Nina, staring through the window at a brown rat slinking across the floor. 'You will have to block windows.'

'But we don't have curtains or anything. And I can't stop everyone looking out.'

Nina grabbed a nearby floral arrangement – an enormous vase with a huge spray of dried pampas grass and yellow chrysanthemums – and thumped it down onto the windowsill.

'There. No one will bloody see through that!' she said triumphantly and made her way down the room, picking up a vase from each table and shoving it onto the window ledges. The large arrangements almost completely blocked out the view.

Colin the pest man arrived and parked discreetly around the back so no one could see the enormous pest control company decal across his van. He flung open the shed door and surveyed the twisting mass of panicking rats. There seemed to be more than ever.

'Aye,' he said phlegmatically, scratching his head. 'That's a fair few you've got there.' He started poking around in the dusty feed sacks, making more rats skitter wildly across the floor. 'They must be coming in from somewhere nearby.'

Colin slowly worked his way around the building, blocking up rat holes, laying poison and replenishing the traps. I followed him, hopping from foot to foot and asking every five minutes how long it might take to finish.

'It'll take as long as it takes,' he said. 'I need to block all of 'em holes, otherwise they'll just come straight back.' He showed me the huge piles of rubble that lay beside each rat tunnel, dug straight down into the foundations of the building.

'They're destructive little beasties,' he said, his eyes gleaming as he flashed his torch into one of the holes. 'Will eat anything. Even each other. There'll be a right little colony under the floors somewhere. Hundreds of nests.'

I straightened up and decided it was time to go and greet the arriving wedding guests. Fortunately, with the huge floral arrangements and the general jollity of the crowd, nobody seemed to notice all the surreptitious activity in the barn outside.

After serving the dessert, I was walking down the bottom passageway with an armful of dirty pudding plates when a little girl, dressed in a frothy concoction of pale pink lace, suddenly appeared around the corner and came to a halt.

I stopped. She stood twisting her petticoat in her fingers.

'There's a big mouse out there,' she said bluntly whilst pointing to the door into the car park.

Oh god, I thought.

I tried to step past her without dropping the plates.

'It was probably just a cute little squirrel,' I said reassuringly.

She thought for a moment. 'Nope. It was big. And brown.'

She rubbed one shoe against her white, lacy tights.

'And it had huge, big teef. Like this.'

She glanced at me from underneath her eyebrows and lifted both hands up in pretend claws, furrowing her forehead and baring her tiny teeth in a grimace to show me what she meant.

Good god, she was terrifying.

I managed to sidle past her to dump my load of dishes into the kitchen.

'IT'S PROBABLY STILL THERE!' she shouted after me as I retreated quickly down the corridor, 'WITH ITS MUM AND DAD!' I hid for a while next to the dishwasher until she disappeared back into the party upstairs.

That was the only rat sighting we had all night, although I took an umbrella out with me to see the wedding guests onto their coach, ready to use it like a cricket bat to launch any encroaching rodents over the wall before anyone saw.

After Colin's ministrations, the High House 'rat plague' quickly diminished so that I could remove the huge flower arrangements from the windows and open the shed door without expecting a wave of toothy rodents.

Colin told us that the mild weather and wet conditions had led to an explosion in the local rat population, and as some farms weren't using effective pest control, there was a carpet of rats moving across the countryside, competing for food and expanding at an alarming rate. It all sounds rather Armageddon-ish, but thanks to regular visits by Colin and his trusty little black boxes we managed to avoid any further ratty plagues. But to this day, I always flick on the shed lights and wait for a second before opening the door, just in case something big with 'huge teef' is waiting for me on the other side.

*

As well as the lovely Nina, the rest of my family become employees in the business. My dad became beer salesman and payment collector, Mum helped in the kitchen and I oversaw the office whilst Steve was busy in the brew hall downstairs. Working with family was an eye-opener. We often fell out, or became irritated in the heat of the moment, but any arguments were quickly forgotten as we settled down together, and everyone learnt their roles and knuckled down to the job. Steve started off brewing three core beers: Nel's Best (named after our farm dog, Nel), Auld Hemp (named after the very first sheepdog in the county) and Matfen Magic (named after the Bronze Age stone in a nearby village). Steve only used Northumbrian malt and English hops, and with his creativity and attention to detail our ales grew in popularity. Our delivery vans would roll out three times a week, with a route up into the wilds of Allendale and beyond, one down to the pubs along the Tyneside Coast and a route into the depths of Durham. We hired two dray drivers, entered competitions and started to gain a small group of fans. I was even pressed into doing brewery tours for visitors: I would take small groups around the fermenting rooms and the conditioning cellar and up into the bar where we would sip ale samples while I explained the different hops and malt. This proved to be a bit of a challenge since, as previously mentioned, my palate is firmly slanted towards sweet drinks and my usual tipples are sugary cocktails. I despise beer, so it was difficult to appear enthused and knowledgeable whilst everything I sipped tasted just like burnt water.

The beer tastings weren't the only hurdle. Whilst I could manage cups of tea and slices of cake, I have never been a confident cook, and it was nerve-racking when we introduced a bigger menu with cooked lunches and evening meals. The day we had our first group booking in for dinner I was so nervous I dropped an entire Tupperware box of tomato soup down the back of the fridge, splashing one side of the little kitchen with huge gobs of orange-red goop. On the same night, due to my lack of foresight, I also discovered we didn't have enough hobs to cook all the vegetables for the planned beef casserole, so Mum, clutching a massive pan of raw carrots, had to wobble out in the dark to use the cooker in our farm cottage. It was pitch black, and on her journey she stepped in a massive pothole on the drive and dropped half the carrots, which gave me the giggles. I had to crouch behind the cake stand with my apron shoved into my mouth to stop the guests from hearing my chuckles. We weren't exactly professional, but the evening went off okay, although there were a few queries about the lack of soup as a starter choice. We eventually hired a professional chef, but remembering those first days brings back a feeling of hysterics and laughter and I'm amazed we managed to cope without any major complaints.

We garnered a fair bit of attention as a new rural-based business and had various visits from EU representatives who wanted to know how we'd spent their money 'regenerating the countryside'. We were eventually contacted by the people surrounding Prince Charles to be involved in a 'meet and greet' that he was carrying out as part of a tour into the wilds of Northumberland. Charles has always tried to champion farming businesses and we were told that we could meet him at a local pub in Greenhead and give him a brief tasting session of our three core beers. The closest my family has ever been to royalty was when Dad showed Princess Anne around a local sewage station, so everyone was very excited about this idea. To say we were nervous was an understatement. Steve had dressed himself in his best tweeds and I had squeezed myself into a little black dress and teetered on a pair of very high heels. At the pub I was provided with a tall stool to perch on whilst a very posh-sounding gentleman briefed us on what we could expect. It was all timed to the second and typed out on little bits of paper: Charles and Camilla would arrive by helicopter in the pub garden, and there they would be introduced to the landlord before eventually coming to the bar and chatting to us about our beers. We were told to wait to be spoken to and then address the royal couple as 'Your Royal Highnesses' in reply, which was rather different to my usual conversational style of launching into a loud discussion peppered with a few of my best jokes. I surreptitiously practised my wobbly curtsey, my heels catching on the carpet, and watched the pub lounge fill with press photographers. Eventually we heard the deep thrum-thrum of a helicopter and after a moment, a wedge of burly Secret Service agents appeared around the corner escorting a bundled-up Prince Charles and Camilla and accompanied by the nervous-looking pub landlord. My first impression was that Prince Charles was surprisingly short. And the second was that his tweed overcoat was beautifully made and probably cost more than our annual farm income. Prince Charles turned to Steve and smiled, shaking his hand and uttering a plummy 'Good morning.' All the camera flash bulbs seemed to explode in one moment, and Steve was completely flummoxed by the sudden blast of light and noise. He managed to dip his head in the required bow, but instead of anything sensible, stuttered out a nervous 'Alright?' as a half-hearted greeting. I wobbled off my stool into a shaky curtsey and shook Camilla's proffered hand but promptly had to sit back down again as my knees gave out.

The photographers were urging Charles to pull a pint so after a bit of chat about the beers he started to haul on the pump, causing half a pint of foam to cascade into the glass. I have absolutely no idea what I talked about to the Duchess of Cornwall while all this was going on. It could have been anything: the fat pony, the brewery, my kids, my latest operatic struggles with constipation. I really hope it wasn't, but all I remember is that my voice was several octaves higher than normal, and I did a lot of nervous, high-pitched laughing. Poor Camilla. After a few minutes the royal couple were ushered back out into the pub beer garden and I sighed gratefully as I slid off my heels and covertly undid the waistband of my tight dress so that I could sit on a bar stool and chat to the photographers whilst listening to the 'whump, whump' of the helicopter blades as the royal couple were whisked off for their next appointment.

*

As High House Farm became better known, we started attracting larger groups and coach tours began using our tearoom as a useful stopping point as we were about halfway between the Roman Army Museum and Vindolanda and the more down-to-earth delights of the Metro Centre. The car park was big enough for the huge buses to turn around, and we were popular as the men in the party could have a brewery tour and the ladies would flock into the tearoom for scones and a drink.

Our first coach party arrived in an enormous, very expensive-looking bus that had tinted windows and a disabled lift. It juddered to a stop in our car park, taking up half the spaces, and the coach driver hopped down, dressed in a smart shirt and tie. The clutch of elderly ladies and gentlemen had already spent the morning at the nearby Hadrian's Wall and had planned to stop at the tearoom for lunch (everyone had already booked a 'Drayman's platter' and a pot of tea) and we'd anticipated a quiet afternoon of genteel conversation and peaceful chatter. But this notion was rudely punctured when the first lady raced off the bus, through the door and into our toilets.

'Her stomach's upset,' said the coach driver with a bored shrug. 'Been in the bog since we left this morning.'

The rest of the group, some with sticks and some with wheeled walkers, hobbled out of the bus and I was kept busy in the tearoom operating our own platform lift and taking drink orders. One miniscule woman, with hair like a puff of white dandelion, gamely shuffled along the floor supporting herself with two crutches. 'This better be bloody worth it,' she muttered as she passed by.

We slogged on, making millions of cups of tea, handing out the lunches and rushing around making sure everyone was sitting comfortably. The whole group sat munching happily until the cheerful hum of chatter was disturbed by a sudden flurry of raised voices at the far end of the tearoom. A red-faced man was on his feet shouting at an icy-looking Nina and shaking his finger an inch from her nose.

It turned out the argument was about cheese. The customer was convinced that the panini he'd ordered (bacon and Northumbrian cheese) didn't actually contain any Northumbrian cheese at all, and as far as I could understand from his beetroot-red face and the words emerging from his tiny, pursed mouth was instead 'crap from the local Tesco'.

Nina was becoming more and more upset. 'It is from here. Look!' she snapped in exasperation, lifting up the wedge of local cheese in her fingers and waving it under the man's nose. I stepped between them and grabbed the slice of cheese from Nina's hand.

She huffed in fury. 'He says it is not real. But it is. I take it out of the box myself.' Adding a long string of mumbled Ukrainian curses, she hissed in a vehement mutter under her breath.

The visitor was now maroon and explosively stabbing his finger into the air.

'This is *not* local cheese!' he bellowed, prodding the sad, melted remains of the panini on his plate. 'It's bright yellow!' He looked around at the others on his table, whilst they stared at him with open mouths.

I shooed a very sweary Nina into the other room, brought the wrapper from the wheel of Northumbrian Nettle Cheese and plonked it on his plate. I could hear Nina banging plates and cups together on another table next door and muttering icily to herself. (Note: Nina's favourite curses were 'Sraka motyka!' – 'Arsehole!' and the rather wonderful 'Shchob u tebe mizh nih vidsokhlo!' – 'May your balls dry out!')

When faced with actual proof that we hadn't committed cheese fraud, the maroon man subsided into his seat, still grumbling about trading standards and labelling regulations. There was no apology, and I couldn't help noticing that he ate the rest of his panini without any more complaint.

After everyone had calmed down and finished their lunches, it was time to gather the group together for my brewery tour. I carefully helped the motley collection of elderly individuals across the malt loft, positioned them in front of the railing and launched into my opening paragraph with a cheery 'Welcome to High House Farm Brewery!'

I was just droning on about the different flavours and aromas of hops when Nina's face appeared at the window, her eyebrows bouncing up and down in agitation.

'What?!' I hissed.

'Problem at till!'

I sighed, made my excuses and slipped through the tearoom door, leaving Steve, knee deep in the copper below, to explain the rest of the brewing process.

When I appeared at the till, Nina darkly pointed out a very frail-looking chap in a wheelchair with a tartan rug over his knees.

'Him. He's taking with no money. Stealing things.'

'But he's ancient!' I whispered. 'He doesn't look well enough to even sit up.'

'Watch him,' she hissed into my ear as his carer wheeled him past the shop shelves piled high with soap and beer shampoo. As I watched, I saw his left hand whip out, grab a bar of soap and shove it under his lap blanket.

'Oh god,' I muttered, 'he bloody *is* too.'

This group tour was just becoming more bizarre by the minute. I squared my shoulders and, feeling dreadful, stopped in front of the wheelchair man and insisted on his carer looking underneath his blanket. His skinny knees were piled high with snaffled brewery pens, bottle tops, magnets, bookmarkers, rubbers, soaps and beer-flavoured lip balms. The old man's hands shook as I scooped up the pile of items and dumped them onto the shop counter.

The lady pushing his wheelchair was bright red with embarrassment. 'I am *so* sorry,' she whispered, 'he can't seem to stop himself. He keeps nicking things everywhere we go, and I keep having to check his pockets. It's so embarrassing.' Reassuring her that there was no harm done, we put the bits and bobs back on the shelves and I went to finish off my tour.

Finally, the cups of tea and halves of real ale were all finished, and Nina and I stood at the top of the steps to watch the tour group climb arthritically up the bus stairs and we waved a relieved goodbye as the doors hissed closed. Everyone slumped against the wall of the tearoom, surveying the remains of twenty-two Drayman's platters, half-drunk pints and many pots of cold tea.

Steve popped his head out of the brewing hall and called me over. 'Come and see the loos,' he whispered, his face white as chalk. I followed him downstairs, peeked round the door of the ladies and was met with a scene of devastation. Everything had gone: the soap, air freshener and even the soap holder, all the paper towels, a decorative pot plant and, worst of all, every single bog roll. The lady with the upset stomach had also left her mark, liberally spattering the pan, the walls and the floor in front of the toilet. I had to go and stand in the fresh air until I felt less green.

Upstairs Nina was staring into the tip jar by the till. 'Those buggers,' she said furiously, 'those buggers have taken our money!' She showed me the completely empty pot. 'It must have been that old man.'

'I think it could have been any of them,' I said as I wearily picked up the bucket and mop from behind the till and started down the stairs. 'They've cleaned us out of all the soap and loo roll too.'

Nina's eyebrows drew down into a furious, black line. 'Thieving buggers,' she hissed as she turned to pick up the bleach and sponge from under the sink.

Aside from the odd unpleasant one, most of the visitors to our tearoom were wonderful. They loved the chance to visit a working farm and to walk along the public footpath that ran through our fields and ten-acre wood. They loved seeing our motley collection of farm animals, including our Shetland pony Candy, our pet lambs and our friendly ewes. We started to receive visitors from nearby Newcastle, especially after we were featured in the local newspaper as a 'must-see' tourist destination. Sometimes there was a collision of cultures between visitors who typified the cool, urban city dweller, and those of us who had more of a rural bumpkin outlook.

The producers of a very popular Newcastle-based reality show got in touch with us to ask if they could film a segment of their show at the brewery. Our farm buildings were apparently perfect as a backdrop to a baby gender reveal for their main celebrity couple, the leading man and his pregnant girlfriend. The producers rang to tell me that they'd organised a motorbike to drive through the farmyard and up a ramp into a haybarn where the rider would press a special button so that the motorbike exhaust would belch out pink or blue smoke, thereby letting the happy onlookers know whether the couple was expecting a boy or a girl.

On a dark winter's day, I watched interestedly as a group of perma-tanned twenty-somethings stepped out of their cars and gingerly tiptoed across the muddy car park. Every single bloke was wearing box-fresh white trainers with skinny jogging bottoms and a tight V-neck jumper showing a few inches of freshly waxed chest. I had been worming sheep, so I was dressed in my wellies, an unfortunate bobble hat and a long, brown coat which stretched down to my shins with some really dodgy-looking stains across the front. I was also limping a bit as I'd hurt my knee that morning when I swung one leg over the quad bike, which added a little extra to my appearance. The leading man was absolutely gigantic, towering above my head, with enormous shoulders, tattoos crawling up his arms and a neck and biceps like the Hulk. He flashed his blinding-white teeth and gently shook my hand with his enormous shovel-like paws. His girlfriend was equally amazing with layers of thick makeup; huge, fluttery false eyelashes; matching bleached teeth and skinny jeans tucked into five-inch-heeled boots.

The filming started and the motorcycle rider manoeuvred his bike into place, dropped his visor and started revving the engine. Suddenly he opened the throttle and roared into the haybarn, sliding his bike to a stop next to a rent-a-crowd of clapping onlookers who squealed with excitement when the bike's exhaust let out a wet fart of blue smoke which drifted across the barn and into the faces of the cameramen. 'Cut!' yelled the director as everyone ran busily around to set up for the next shot. I was transfixed. I'd never been so close to actual celebrities before (apart from Prince Charles, and he'd been disappointingly short). A very young assistant came across and asked me if I'd be happy to be in some background shots, to show some 'local colour'. Of course I would. I was madly enamoured of the whole filming process, so I allowed myself to be manoeuvred to stand beside a hedge for a panning shot across the farm entrance.

'Just look natural!' shouted the director, as I stood shivering in my filthy coat. I'm not sure what he meant but for some godforsaken reason I decided that it would look extra casual if I started collecting firewood and pinecones from the grass. At the shout of 'Action!' I stooped and started to collect an armful of twigs whilst smiling inanely at the ground in front of my feet. I must have looked like a strange rural halfwit, collecting firewood for a flickering peat fire back at my filthy one-roomed cottage. After that they asked me to lead a bored and filthy Candy across the farmyard for a background shot. By this time my knee was really hurting, and I started to limp wildly, my bobble hat hanging over one eye as I hauled the muddy pony across the cobbles, muttering inanely into one fluffy ear. I have no doubt that the producer probably took one look at the film of my appearance and decided that the rural inhabitants were just too weird to be added into the show. My bits of film were never broadcast on TV, but somewhere in the depths of the BBC archives will be some shots of a simple peasant woman lurching across the screen whilst chatting about the weather to her overweight pet pony.

*

Weddings became very popular, especially as we realised we were much cheaper than the nearby five-star hotel's much pricier package. From 2006 to 2009 enquiries flooded in, and I did venue tours for hundreds of prospective couples, created individual menu plans and canape suggestions and gathered together ideas on florists, wedding singers and dance bands. The old Dutch barn was turned into a magical wonderland of flowers and lights whilst the outside stairs and tearoom were festooned with streamers of greenery and pretty silk flowers. In the local wedding magazines we were billed as 'a unique and unusual wedding venue with bags of rural charm and an enchanting countryside backdrop'.

We attracted all sorts of couples and we had every single type of wedding you could think of: same-sex civil ceremonies; pagan couples with complicated rituals; luxury receptions with miles of expensive hothouse flowers and (hired) designer furniture and light fittings. Our memorable ceremonies include a bizarre lesbian cowboy reception with the registrar wearing a silver sheriff badge, toy guns and ten-gallon hats, and a Star Wars ceremony complete with everyone in costume, a massive spaceship cake and a real live Wookiee. The creativity and uniqueness of some of the wedding parties was incredible.

One of my favourites was the gay pagan couple who turned up on a decorated float in bright red suits alongside their high priestess (in a crimson velvet robe) and an enormous ceremonial sword. After a registry wedding and then a handfasting ceremony filled with banging drums, flutes and singing, the newly married pair asked if they could walk down to our wood and spend some time with an oak, an ash and a rowan tree.

'Take your pick,' I said. 'What are you going to do down there?'

There was a nonchalant explanation of 'communing and nature worship' and to this day I'm still not sure what they did in the forest, but I really hope they enjoyed themselves. They were great fun: kind, open-minded and best of all, extremely generous tippers.

Our animals on the farm entered enthusiastically into the wedding ceremonies and celebrations. At that time, we had a gaggle of free-range chickens and ducks who, during the day, were let out of their houses to bobble around the farm, pecking at worms and bugs and laying eggs in various hard-to-reach places.

Marjorie the Fourth was one of the biggest hens, with smart red feathers and a very fetching apricot-coloured, fluffy bottom. She was very friendly and used to take her entertainment role seriously, stamping determinedly up to wedding parties and clucking loudly at them in the hope of being fed a few handfuls of canapes. The weddings were solemnised in our haybarn, which meant Marjorie could also join in the civil ceremonies, sometimes walking down the aisle ahead of the bride, drawing gasps and giggles from the guests or trailing behind the wedding party to become an uninvited additional bridesmaid. At first I tried apologising and shooing Marjorie away, but I soon realised that our wedding parties loved her joining in, and she became a star in her own right. She also managed to insert herself into many wedding photos, usually pecking around at the feet of the bride and groom or carefully inspecting the photographer's camera. I would imagine that many wedding photos involved a close-up of Marjorie's unblinking, black button eye.

Candy – our small, fat Shetland pony – was also involved in weddings, either as a photographer's prop (she was good at standing in the back of wedding photos although it was difficult to keep her away from the wedding bouquet) or as entertainment for the younger guests. Candy was bombproof, and she didn't care which small person was sat on her back as long as she was allowed to eat the grass. One bride asked if I could lead her through the marquee as a treat for her tiny bridesmaids, so after the wedding breakfast had been eaten, and speeches and toasts had been made,

I carefully led Candy, who had been brushed for the occasion and was wearing a pink ribbon plaited into her forelock, across the dance floor to the top wedding table.

Candy looked around with interest, nosing me to see if I had any treats in my pocket. She knew that being brushed and led around on a bit of string usually meant that she would be fed something tasty, so she busily inspected the wedding table, attempting to chomp the pieces of confetti. The wedding band started up and the small pony didn't bat an eyelid, ignoring the music and DJ as she tipped one rear hoof to rest her foot calmly against the floor. After the tiny bridesmaids had patted her nose and stroked her neck, I led Candy over the dance floor to the marquee exit. As we passed the remains of the buffet table, the little pony suddenly made a dive to one side, sticking her nose into the depths of a salad bowl whilst I tugged ineffectually on her lead rope. Candy was determined to eat her fill, and before I managed to haul her away from the table, she'd scoffed three vol au vents and some leftover wedding cake. Hoping no one had noticed, I dragged her out of the tent, and she pranced behind me back to her field, immensely pleased with herself and licking her chops thoughtfully. Marquee weddings were a free-for-all for the chickens as well, and whilst they clustered under the buffet table eating dropped lettuce the bolder Marjorie and her friend Ethnee could be spotted at the edges of the dance floor, pecking for crumbs among the dancers or harassing small bridesmaids and elderly aunts into giving them a crisp. After the buffet incident, Candy was always very interested in subsequent wedding marquees and if I led her past a large tent, she'd peer hopefully into its depths on the off chance that she'd be allowed another go at the wedding breakfast. Our farm cat and collie dog also tried their best to be included, and the morning after a big wedding, I'd spot Cyril and Nel the dog licking up the grease from the stones under the hog roast spit.

Merging two families is often a tense affair, especially where difficult personalities are concerned. I spent a lot of time calming down irate in-laws or smoothing over family arguments, and it was surprising how many times I had to coax a crying bride or wobbly groom down the aisle. I was even more surprised at how many times we needed to call an ambulance to a wedding to ferry an injured wedding guest to the local A&E. It always seemed to be tiny, ancient grandmothers who would break an ankle or a hip through overenthusiastic dancing or drinking too many glasses of wine and tripping over a step. One weekend we had to call the paramedics out three times in a row to weddings held on the Friday, Saturday and Sunday nights. The first call-out was for a ninety-year-old grandmother who had chest pains, the second was to deal with a very drunk uncle who fell down a step and the third was a gentleman who gave himself a hernia by turning a somersault on the dance floor. We became good friends with some of the ambulance staff.

One lovely bridal party had spent a fortune on renting tented wedding 'tipis' to be built in our car park. These are enormous canvas wigwams that are connected to each other with yards of fabric and wooden poles and can hold a large number of guests. This family had spared no expense. One tipi served as a dance floor (with a proper sprung wooden floor) with DJ decks, lights and speakers, and the other side held a fire pit, tables and chairs to seat over fifty guests, all festooned with miles and miles of fairy lights, greenery and flowers.

The family were from London but had chosen to get married in Northumberland as the cost of wedding packages was a tenth of those in the capital. Guests arrived on a red London bus (great excitement from my kids as it drove into the car park) and sported huge hats and long, floaty dresses. The bride arrived in a vintage sports car and climbed out of her seat looking pale and jittery. She looked gorgeous. She wore a vintage lace dress with a flower crown and a delicately embroidered veil that floated onto the floor behind. By the time she reached the brewery door (we brought all brides in through the rear door so that no one saw them before they walked down the aisle) she was green with nerves.

'I can't do it.' She sank down to her knees in the downstairs corridor as her dress pooled in thick folds of satin around her feet and ankles. She started to cry as I rushed to get tissues and a glass of champagne, coaxing her to have a sip of her drink and to tell me what was going on.

'I just can't do it. I can't get married and I can't wear this stupid dress in front of everyone out there.' The bride flicked her hand at the froths of creamy lace that skimmed her tiny waist and fell to the floor. 'I'm so nervous and I just want everyone to go *home*!'

Her nervousness at being the centre of attention made me remember my own wedding day. 'Just take deep breaths,' I said. 'Take as long as you like.' Her dad started patting her shoulder awkwardly, muttering various platitudes as his daughter gulped miserably into her handkerchief. Her mum rushed in and took her off to sit in the conditioning room – I peeked in ten minutes later and the bride looked a little better, still very pale but looking a little more cheerful as she sat next to her mum on an empty cask of ale.

Eventually she *did* walk down the aisle and the ceremony carried on without a hitch. I later spotted the bride dancing with her groom in the tipi – they were swaying happily and whispering romantically into each other's ears.

My favourite part of the whole wedding business was seeing the excitement of an engaged couple as they discussed their vision of their perfect day. It made me remember how I felt as Steve's fiancée, flipping through dress brochures and wedding menus, feeling the butterflies in my stomach when I started pulling together guest lists and arranging the tiny touches that made my day so meaningful and memorable.

Our popularity in the wedding business grew tenfold between 2002 and 2009 and the phone wouldn't stop ringing with new enquiries. The brewery was also very busy, and we now had a stable of over eight different ales (all named after our farm animals with such gems as Cyril the Magnificent after my favourite cat and Ferocious Fred, named for our scary bull). The tearoom was ticking over nicely and there were ten staff members on the books.

And then I became pregnant with my first baby. I was excited and happy and nervous, and Steve was delighted with the thought of becoming a dad. After the first thrilling month I was also absolutely floored with morning sickness (or more like all-day-and-night sickness) which seemed to morph into deep and unrelenting fatigue and anxiety. The smell of the boiling malt made me incredibly nauseous, and I had to delegate all the wedding show rounds as I spent most of the day in bed or kneeling in front of the loo. The midwives and health visitors stepped in, changing my antidepressants that I'd been taking on and off since my twenties, giving me some anti-sickness pills and seeing me weekly as I sat in their consulting rooms, sobbing into a hankie and trying not to throw up in their laps.

Steve was trying to look after me, juggle farm work, brew three times a week and help in the office. During lambing or harvest he was literally on his feet almost twenty-four hours a day, catching a quick nap here and there as he struggled to bring in the wheat or catch up with the demand for different ales.

I staggered up and down the stairs, my bump and sickness making it more and more difficult to oversee weddings into the wee small hours, or even help the waitresses with clearing plates or picking up glasses. I couldn't afford any maternity leave, so I worked right up until the birth, my back hurting and legs aching. Lucy was born two weeks early, after a thirty-six hour labour and an emergency C section. Unfortunately, I almost immediately had to go back into hospital for a blood transfusion and a few days of antibiotics through a drip due to a nasty infection. I was a besotted new mum but also teeth-clenchingly anxious and could barely drag myself out of bed. Fortunately, due to a concerned GP, I was quickly diagnosed with a case of postnatal depression, and the NHS swung into action, delivering sympathetic counsellors and health visitors to my door and prescribing a myriad of different drugs. I spent the first few months as a new mum trying to look after my baby, taking advantage of any gaps between wedding venue tours to breastfeed in the office, carrying my precious little girl around in a papoose as I blearily organised weddings, answered emails and phone calls and dealt with staffing issues.

Halfway through 2009 we made the momentous decision to hand over the wedding business to a new family. It became impossible to stay up until 2 a.m., trying to hoick tipsy wedding guests into their coaches or to soothe a bawling, red-faced Lucy whilst I answered the phone or printed out dray runs and cask labels. It was definitely time to rethink and move on. We decided that instead of running everything ourselves we would become landlords and rent the buildings and fittings to a new owner. Stepping back would mean that the rent would cover our loans and mortgage and more importantly free up enough time so that Steve could run the farm, and I could finally enjoy some time with my new baby. It was a real relief to let someone else take control, although explaining to people why we gave up running the business was tricky as I still felt ashamed about my struggles with anxiety and depression. Luckily a local family had heard that we were hoping to sell and quickly stepped in to carry on the business, and under their guidance, High House went from strength to strength. Many hundreds of couples were married in our beautiful haybarn and enjoyed their special day on our farm.

And even though we've stepped back, living in the farmhouse means that I still get to experience the fun and excitement of weddings, and the happiness reminds me so much of my own. Stepping outside our front door on a Saturday means that I regularly bump into bridal couples posing for photographs next to our overly interested pony, or see the arrival of a bevy of excited, chattering bridesmaids in a vintage campervan. I still find it an absolute privilege to share our farm and buildings with these new couples and have a glimpse into the fun and laughter of their special day.

CHAPTER FIVE

Sheep-Based Misadventures

To my mind, the life of a lamb is no less precious than that of a human being.

Mahatma Gandhi

For god's sake don't stand there – stand THERE!

Steve Urwin

My bedroom window faces our twenty-four-acre back field, and I can see all the way to the oak and ash trees on the fence line and beyond to the villages of Matfen and Ingoe. I found an old map of the farm in the Northumberland Archives, wonderfully hand drawn in scratchy pen and pencil, showing all the boundaries and field names. The map is dated 1830 and carefully traces the contours of this field, which used to be divided into two and named rather wonderfully Lumpy Lands and Henry's Tack. Lumpy Lands is so called due to the rig and furrow that criss-crosses the turf as well as huge bumpy hollows which look as if previous farmers have dug for sand or building materials. Henry's

Tack is named after a long-dead Henry who wintered his sheep on the sheltered ten acres to the west of the back field. 'Tack' is an old word meaning a place rented for winter grazing a sheep flock.

Sheep have been part of High House Farm for hundreds and hundreds of years. Our old farm buildings have many traces of ancient sheep folds and sheds. In my diggings in the Northumberland Archives, I've found reports from the 1850s written by previous farm managers that have screeds of careful lists of numbers of sheep and how during the winter, due to a lack of grazing, they fed their flock turnips and 'steamed roots' that were prepared in the steam-driven engine on the farm.

I'm glad we can continue the tradition of keeping sheep at High House. Our current flock numbers one hundred-and-fifty ewes, two hundred lambs and five tups (male sheep). The girls are a mix of Beltex, Texel, North of England Mules and Suffolks, whilst our boys are all pedigree Texels and Beltex. These breeds produce chunky, well-muscled and meaty lambs, perfect to sell at the local mart for buyers in the UK and across Europe. We have previously owned Cheviot sheep (pronounced 'Chivvots') but no longer, as we spent most of our time collecting them from next door's farm as they had a very nonchalant approach to fencing and seemed to enjoy using our stone walls for mountaineering practice.

The sheepy year follows a well-worn timetable, from tupping in November to lambing in March and April, and weaning or 'spayning' in August. I like to think that generations and generations of sheep farmers must have followed the same general calendar with only small tweaks or changes. In November our five tups (Thrusty Clappernuts, Randy Jackhammer, Batty Blueballs, Ron Bangerplums and Dave) are strapped into harnesses and their chests are liberally smeared in oily, yellow chalk. Each tup is given fifty ewes and off they gallop for twenty-one days to mate with their ladies and in the process daub yellow all over the ewes' behinds. After twenty-one days the boys are smeared in orange, a slightly darker colour, and shooed back into their fields for more frolicking. Tups will only mount a ewe if she's still in season (and therefore not pregnant) so by checking the colours on the girls' rumps we can work out which ewes are either not mated or when approximately their lambs will be due. We change the colours every twenty-one days (the lengths of a ewe's breeding cycle) from light to dark chalk – yellow, orange, green, red and blue – so that it easily shows up on the ewe's woolly backside. This method means that we can work out exactly which ewes are fertile and which are not, and when they're going to lamb. The science behind sheep breeding is complex and detailed, and it took me a very long time to piece it together and fully understand. Steve has a very clever computer system to work out the technicalities and which breed traits we want to avoid and which we try to encourage to improve our lamb health and weight.

It feels a bit awkward staring at our tups 'in action' but we need to occasionally check that they haven't forgotten what to do or lost interest and that all their bits are working. Shouting encouragement out of the window at a lugubrious Thrusty Clappernuts whilst he's balanced on top of a bored ewe becomes second nature, although it does tend to be a bit of a conversation stopper when friends are visiting.

Lambing is the *main* event of the year, and it is literally all hands on deck. When the kids were small, it was very difficult managing the lambing workload and looking after them and keeping them safe. I used to rely heavily on my parents for babysitting and housework as it was impossible trying to feed the children lunch or put them down for a nap when I was needed back out on the farm to help a ewe give birth or to feed a sickly newborn lamb. As the children have grown bigger and older, they have become a great help during the busiest time of the year: helping fetch and carry, driving the quad bike to check on the older ewes and lambs and even helping catch pregnant ewes and lambing themselves. Lambing starts on 3 April each year, and every twelve months I give myself a good talking to about toughening up. I have a marshmallowy soft spot for those lambs that struggle, that are born a bit different and can't nurse like ordinary babies. Any catastrophic scenario pulls at my heartstrings: from lambs that have been abandoned by their uninterested mothers to tiny scraps that won't suck or are too weak to stand. You'll find me in the lambing shed, surrounded by ewes and lambs and gently propping up some titchy, woolly dot so that I can siphon small dribbles of warm colostrum into its stomach. I'm sure other farmers are much more hard-nosed and euthanise babies that they don't think will survive, but as long as they're not in pain I'll always give an animal a chance. Steve is a rough, tough Northumbrian farmer on the surface, but underneath his hard, grumpy exterior he's as 'soft as clarts' as they say around here. Of course, he doesn't want anyone to know, and he's pleased that he can blame any waifs and strays on me or the kids.

He's keen to point out to visitors that he's not the one looking after some wobbly and pathetic lamb. 'The kids are looking after that,' he'll say. Or 'The wife deals with all the runts, I've not got the patience.'

But then the feeble lamb becomes a loved pet sheep and in a few months can be found round the back of the farm, well away from the road and curious neighbours, in a paddock alongside a resigned-looking Shetland pony. No mart will sell a disabled or sick sheep so anything that doesn't meet their rigorous standards either needs to be humanely euthanised at birth, becomes a pet sheep for the rest of its life or has to be put down (at expense) by the vet.

Blind Sheep was my first 'rescue' case. He was a cream-coloured lamb with black knees and nose who had eyes that were too small and hadn't developed properly in the womb. He would stand with his massive ears out sideways, looking like a sentient wingnut, twitching at the slightest noise and walking in small circles until he bumped into his mum. Once he got the hang of drinking her milk he grew fast and strong, but out in the fields he would regularly be left behind when she wandered off to graze. Blind Sheep would be snoozing peacefully in the spring sunshine until he realised that he'd been abandoned and no one was near. He'd leap up from a deep sleep, blarting in panic, and lurch off in ever-increasing circles until he could hear his mum's exasperated bleats in return. He grew up to be a very friendly (if slightly special) sheep, and once castrated and weaned from his mum we used him as a companion to anything that needed a bit of love and affection. Blind Sheep loved mothering pet lambs or keeping our ponies company, and he used to snuggle up close to whatever reluctant animal he was with, breathing heavily in their faces in dim adoration. He eventually died of old age but for a long time was a stalwart on the farm, listening carefully, ears stuck out sideways as he followed us around the yard, looking for sheep feed in people's pockets.

Then there was Wobble the lamb. During the 2020 lambing season, Dad and I were sent off to buy some spare lambs from another farm to adopt onto some of our ewes that had miscarried. (We do this sometimes with mothers who abort their babies if they have plenty of milk to spare to feed an orphan lamb.) At a farm in Durham the shepherd showed us a clutch of titchy Mule lambs with fluffy legs and blotchy, black and white faces. We paid for three of the babies (the going rate was a tenner each) and stuck them in the back of the trailer, but just as we were about to pull out of the yard, the shepherd appeared around the trailer door carrying a large, ungainly looking lamb with long legs and a wedge-shaped head.

'D'you want this one? It's a bit shaky maybe, but it's sucked from its mam alright.'

The lamb had a triangle-shaped head, a tiny black nose and a curly coat. The shepherd put it on the floor, and it immediately started blarting whilst its whole body jiggled from side to side in agitation.

Dad and I looked at each other.

'I'll give it to you for a fiver,' wheedled the shepherd.

Dad peeled a five-pound note out of his wallet whilst I wrapped the wobbly lamb up in my coat and sat him on my knee. On the way home he peered out of the window, his head gently quivering against my hand.

'I'll call him Wobble,' I said to Dad, who raised his eyebrows and reached across to give the lamb a brief scratch on his little woolly head.

Back at the farm Steve looked on in disbelief as I guiltily unwrapped Wobble and placed him gently on the stone floor. Wobble sniffed at his feet and then shook his way across the pen to stand hunched up under a warm heat lamp.

'What the hell is *that*?!' said Steve.

'I couldn't leave him behind,' I said in a small voice. 'He'd have been put down.'

'He cost me a fiver,' shouted Dad from the back of the shed.

'A fiver?! They should have paid us to take him. You've been *had*.'

I watched as Wobble shook his way over to his pen mates, his wobbly back end doing a rumba as he nuzzled under their bellies in a futile attempt to find some milk.

'Right,' said Steve, rubbing a hand down his face, 'he's your responsibility and for god's sake don't put him on the internet. I don't want anyone knowing that we're running a sanctuary for weird-looking lambs.'

I published a video of Wobble on Twitter the very next day and the response was immediate. Everyone loved him and his juddery legs and body. I spent ages trying to get him to drink from a spare ewe, but as he got near her udder and smelt the milk he grew more and more excited, and his wobbling became so wild and unhinged he found it impossible to stay still enough to actually suck. The ewe looked on in horror as Wobble involuntarily tap-danced from one side of the pen to the other and then point blank refused to have anything else to do with him. I shoved the trembling lamb into the pet lamb pen with the rest of the bottle-fed babies. I had to wedge him under my arm when I fed him, to stop the constant trembling so he could suck and swallow the milk, but as he grew in size and confidence the wild shuddering started to decrease, and he began to keep up with his flock mates. The vet was of the opinion that he'd been starved of oxygen during birth which caused some kind of neurological issue.

Wobble was a happy pet lamb and hit all his milestones, being weaned from powdered milk to lamb feed, and then leaving the pet lamb pen with the others to go and live in the sheltered paddock next to our house. His wobbling became less and less, and it was only noticeable if he grew excited during feeding time or if we had to move him from one place to another. Then Wobble would shiver and shake and sometimes fall over with a crash if there was too much stimulation. He looked a bit embarrassed to be suddenly face down in the grass but would pick himself up and trot shakily after his flock mates, usually with his fleece covered in mud. I'm afraid this story doesn't have a happy ending, as unfortunately Wobble didn't make it to adulthood – we found him one day curled up under the trailer, where he had peacefully lain down and died. We couldn't work out what had killed him, as he was too big for a fox or crow attack, but eventually decided it must have been a stroke or some kind of catastrophic brain failure linked to his neurological issues.

It's not all doom and gloom though. Among the hundreds of sheep that have lived on our farm there have been some definite favourites. Scabby Ewe (affectionately known as Scabs or the Scabster or the Scabmeister) is one of them.

Steve was always trying to improve our flock by adding new ewes or tups. A few years ago, we scraped together the money to buy twenty new in-lamb ewes from a wild and woolly farm up in the Coquet Valley. Steve drove up one day in spring to have a good look over the sheep that were for sale. He chose a group of chunky ewes, strong and healthy with good teeth and working udders, all in lamb to a quality Suffolk tup.

We organised a livestock haulier for the following day, and everyone stood around excitedly as it arrived back at High House. Steve heaved open the back of the lorry and counted twenty bouncy sheep down the trailer ramp and into the sheep shed, which was prepared for their arrival with a bed of deep straw.

Steve was chatting to the lorry driver, and as I started to lift the ramp back into place I caught a flash of movement at the back of the truck. Calling to Steve, I watched in horror as a skeletal sheep emerged from the gloom and started wobbling unsteadily down the ramp, blinking her eyes against the afternoon sun.

It was one of the thinnest sheep I'd ever seen, with visible ribs and pipe-cleaner legs. She didn't seem to be the same breed as the rest of the trailer load; instead of a black face and legs she had a splotchy, brown and white head, big ears that stuck out at right angles, bandy knock-knees and a patchwork of filthy, grey wool over her jutting spine.

We watched in silence, mouths agape, as the skinny ewe tottered down the incline, carefully lowered herself off the bottom of the ramp and lurched into the sheep shed to join her friends.

'What the hell is that?!' said Steve, eventually finding his voice.

The ewe puttered uncertainly around her new shed, bouncing gently off her flock mates. Eventually she found a soft spot in the straw, lay down with a sigh and immediately fell sleep.

To avoid any shouting, I volunteered to ring the sheep salesman to ask him how on earth we'd managed to buy twenty new ewes but end up with this one strange addition.

'Oh bloody hell,' he said, when he answered the phone. 'I must have put her in the wrong batch. She was going to the knacker man at the end of the week.'

'Is she in lamb?' I asked.

'Oh god no. No no no no. Definitely not. No, not been anywhere near a tup. There's something the matter with her. I wouldn't put her in lamb. But don't send her back. I really don't want her. You can have her for nowt.'

The spindly ewe was still asleep when we got back to the shed, and she lay there quietly as Steve carefully checked her over. Her mouth and jaw were very strange.

'She's got a shuttlegob,' he said, moving her mouth gently up and down. 'Her bottom teeth don't meet with the top pad. She won't be able to chew grass. No wonder she's so thin. I'll have to send her to the knacker man. She can hardly stand up…'

The ewe was now lying on her chest, with all four skinny legs tucked underneath her body, gamely sucking on a wisp of hay.

There was no way I was going to pass her on to someone else, or put her down.

'She might be in lamb. Can't we keep her just in case?'

After a long discussion, Steve agreed that I could look after her as long as I promised not to tell any of the rough, tough sheep farmers in Northumberland that we'd turned into an animal rescue operation.

The kids named her 'Scabby', and through trial and error I discovered that she could eat sheep feed by hoovering it into her mouth and sort of sucking it until it was soft enough to swallow. From that day on she ate her body weight in very expensive sheep 'nuts' (pelleted sheep feed) and grew very, very tame.

My son adopted her as his 'most favouritest sheep EVER', and she began to follow us around the farm, poking her nose into coat pockets and horse buckets, inspecting feed troughs and hay bales. Scabby even slowly started to put on weight, her hollow flanks filling out a little and her legs becoming less twig-like.

And she *was* in lamb. The scanner man told us she was carrying a single. Which was just as well, because no matter how much feed we poured into her, she never really put on much weight.

We kept her in the shed, out of the weather for the rest of the year, until she managed to heave out a skinny scrap of a baby with splotchy knees and bat-like ears. Scabby turned out to be a very good mother, and with lots of extra sheep nuts she managed to produce just enough milk to feed her lamb.

Once the lamb had been weaned, Steve put his foot down and took Scabby out of the flock into a small group of 'cast ewes' who were destined to go to the mart sale in the spring. I rebelled and kept secretly liberating her from the shed so she could follow me around and get in the way.

One day I got a call from Steve on my mobile.

'Look, I've got Scabby on the scales, and she actually weighs enough to go to the mart. Can I take her tomorrow?'

The Save Scabby Project swung into action and after a lot of wailing, Steve was forbidden to load her up with the rest of the old ewes. I even sprayed 'Scabby' – with a heart shape – onto her side in blue marker paint so that he'd be too embarrassed to take her into the ring.

So Scabby remains. And I'm not certain that Steve intended to sell her. I don't think he'd dare. Scabby is the High House mascot and spends her time lying under a gorse bush, snoozing in the sunshine. She's at the bottom of the pecking order and spends a lot of time on the outskirts of her flock, but she acts as a sort of unofficial auntie to the newborn lambs and young ewes, gamely bobbling around the fields, inhaling grass blade by single blade while costing an absolute fortune in sheep feed. She's so tame that she has a terrible habit of escaping her field to look for sheepy snacks, squeezing through the hedge to stand by the side of the road and look forlornly at passing cars, hoping they'll stop and offer her a rich tea biscuit. She's well known by our neighbours, and there is an unofficial alert system headed up by our postman. He'll regularly spot Scabby mooching on the grass verge, throw open our front door after a cursory single knock and bellow, 'Your weird sheep is out again!' up the stairs.

Flushed with success at the achievement of adding new ewes (and Scabby) to the flock, the next year we bought nine strapping Suffolk cross ewes with four-week-old lambs at foot. They were lovely, chunky, solid sheep with apricot-coloured fleeces and jet-black faces and legs, and I named one of them Josephine as she had an Empress-like nose and an air of superiority.

They were delivered to the farm and decanted into the small paddock which was knee deep in lush grass. They settled down to graze happily, investigating the fences and hoovering through the abundant feed. I spent a few minutes leaning over the gate, congratulating myself on managing to afford such fine animals and admiring their tightly curled fleeces, Roman noses and strong, stocky black legs.

It's pure superstition, but I should have known that things were going too well and I was becoming too confident. The God of Farming likes to wait for these occasional buoyant bursts to lull you into a false sense of security before delivering a bolt of bad luck which destroys any hints of happiness. That night, the ewes spent several hours pushing at the hedge next to the gate until they'd made a hole big enough to squeeze through the gap and trot out into the road, presumably to enjoy a few hours of freedom. They spent a happy half hour padding down the front yard, inspecting the grass around the brewery entrance and visiting our Shetland pony over the gate to her field. After a couple of hours they grew more adventurous and trotted down the road and wandered into Posh Neighbour's garden.

Josephine and her mates then spent a happy evening sampling his shrubbery until they found a really delicious plant. A fetching shrub with bright yellow and green leaves and pretty, pink flowers. Being Suffolk ewes, they ate it until they were stuffed full, then, with bellies bursting, waddled back into the farmyard and squeezed back into their field.

The next morning, before anyone was out of bed, there was a thunderous knocking on our front door.

I blearily opened the door, and an early morning dog walker thrust her face into the gap.

'There's some sheep in that field there making a really funny noise.' She was pointing urgently towards the paddock and as I poked my head out into the morning air, I could hear a deep groaning noise.

I thanked her and called for Steve, shoving on my boots and racing across to the paddock.

It was like a battlefield. Three of the ewes were flat on their sides, stone dead, with legs outstretched and bodies already beginning to bloat. Their orphaned lambs stood pathetically next to the bodies, nosing hopelessly at their mother's fleeces, whilst the six remaining ewes were scattered in different corners of the field, lying on their sides, frothing at the mouth and emitting awful death-like groans.

I couldn't believe it. Yesterday we'd had nine healthy-looking animals, and now a third were dead and the rest looked as if they weren't far behind. We rang the vet, who came and gave me his hanky to cry into and made his way round the field taking bloods and giving antibiotics and calcium.

'It's a textbook case of poisoning. They've eaten something and it's shutting down their systems.'

He poked around the field looking for some plant or tree that might have caused such a reaction. 'Ah,' he said, noticing the hole in the hedge, 'it looks like they've been for some midnight wanderings…'

I had my head in my hands and Steve kicked out at the fence post, furious at this sudden bad luck.

The vet organised post-mortems for the dead ewes, whilst Steve rang the knacker van to pick up the corpses. The remaining sheep were looking a little better after their injections of antibiotics and one or two were even on their feet, blearily nosing at their lambs as they ducked under their bellies to drink.

We cajoled and half carried the poorly ewes and lambs into the shed and bedded them up on straw in warm pens. During the afternoon, two more died, their tongues lolling out of their mouths and their eyes rolling back into their heads. I gathered up all the orphan lambs and started disinfecting bottles and pulling out bags of powdered lamb milk. The lambs were big and strong and even though they were thirsty, were almost impossible to persuade to drink from a teat. They shoved and twisted, frustrated and afraid, as I tried to make them take a few gulps.

When something goes wrong in farming, it's impossible to hide away as the work still needs to be done and the animals still need looking after. So, we soldiered on, nursing the poorly ewes and trying to feed the enormous lambs every four hours. I was tired and fed up and covered in splashes of sticky bottle milk. Later in the afternoon, Lucy and Ben came home from school and immediately pitched in, mixing milk powder and warm water, and helping to string up heat lamps above the poorly ewes in an attempt to keep them warm. Lucy made dinner that night, realising that Steve and I were pushed to the limit, and after feeding herself and her brother, put the pan of stew into the oven to keep warm until we could eventually leave the sheep.

When I'd finally finished the last feed and could leave the pen to go back to the house, I was almost in tears of gratitude to find the fire lit, a warm plate of stew ready and waiting and Ben busily making me a cup of tea. I sat slumped on the sofa for the rest of that evening, cuddled up with Steve and both children, happy with their company as they chatted away about school and friends, trying their hardest to cheer us both up.

The next morning the vet rang and told us that the post-mortem showed that the ewes had been poisoned by an azalea shrub. With the dead ewes, the cost of powdered milk, the cost of getting the knacker man and the final vet bill we reckoned we'd lost around eight hundred pounds. We weren't insured for 'misadventure' so that money had to be written off – a large amount when you're only just managing to survive as a small family farm. The other ewes slowly got better, although one seemed to have lost her sight and another made a weird 'mrrrrrmmmmmm' noise every time she breathed in. They were never as strong as they were when they first arrived, and at the first chance we sold them at the mart, trying to claw back some of the lost money from their purchase.

But this is how farming goes – no sheep had ever decided to push through our hedges before, and no sheep in my memory had ever decided to taste the pretty azalea or rhododendron shrubs in our neighbour's garden. It was just bad luck.

*

When my first book was published, the lovely marketing team asked me to go on a tour around the best bookshops in the North of England to sign copies of my hardback. It was during lambing season, and I had the brilliant idea of taking a newborn pet lamb with me on my book tour. Everyone would love it! People would flock to the store to stroke the cute lamb (and hopefully to buy my book). How difficult could it be to organise?

I first had an inkling about the problems involved when I spoke to a friend on a nearby farm.

'Sal, you can't just rock up anywhere with a random sheep. You'll get fined or shut down or something,' she said.

'I thought it was just like a dog. If you've got a collar and lead surely you can take it wherever you want?'

'There's all these DEFRA rules,' she replied, 'and what are you going to do when it poos all over the floor? I've never heard of anyone potty training a sheep.'

She had a point, but I couldn't back down. I'd told my publishers that I was going to bring the cutest lamb that Northumberland could provide, and in turn they'd told all of the bookshop managers who were busily adding it to their 'meet the author' events. I could hardly turn up at an excited bookstore alone...

First of all, I had to pick the lamb. A ewe had just had a set of triplets and the third lamb was very, very small, almost a third the size of its two bigger siblings. I'd named the tiny newborn Mabel, and as she'd started to feed on the bottle I decided she'd be perfect for a pet lamb tour. She was no bigger than a puppy, with sticky-out ears, a creamy fleece, a pink nose and black socks, and her huge lugs waggled backwards and forwards when she swallowed a mouthful of milk. Everyone would love her; I just knew it.

The first thing to do was to make her tame enough to wear a harness and lead. Sheeps' first line of defence when meeting something scary is to run away in the opposite direction, and I wanted to ensure that she felt safe with me – her surrogate mother.

I took her everywhere around the farm, at first just in my coat with a blanket wrapped around her bottom in case of accidents. I found a tiny puppy harness and lead on eBay and we practised wearing it whilst she drank her milk, and soon she would skip over in excitement whenever she spotted me bringing it into the pen. Then we began to take walks outside and she grew used to meeting different visitors in the brewery or tearoom. If Mabel felt uncertain about anything she'd dart around my legs to hide behind my knees, or if I was sitting down, bounce up onto my lap and settle down in my arms.

House training didn't go so well. I attempted to guess when she needed a wee and would take her outside, but Mabel never seemed to make the connection between needing the toilet and waiting until she was in the garden. After stepping in yet another small pile of lamb poo and smearing it over my kitchen floor, I decided to experiment with Mabel and nappies. At first, the only size she fitted was newborn, and I managed to keep it on by wrapping it around her bottom and shoving a Babygro over the top. One farmer said I should have cut a hole in the nappy to let her tail pop through, but that just increased the poo/floor ratio.

Steve was eating his tea in the kitchen when Mabel proudly trotted around the corner wearing her new nappy and a fetching Babygro that was decorated with pink teddies.

He paused with his fork halfway to his mouth.

'She can't wear that. People will think you've gone mental,' he said.

Mabel stotted around the kitchen, bouncing on all four legs whilst the nappy started to slip down one leg and a popper came undone on her Babygro. Steve watched me as I picked up a roll of gaffer tape and chased her under the table, trying to grab one of her back legs.

'It's the only way I can stop her pooing on the floor,' I panted, wrestling a happy Mabel onto my knee.

Steve looked on in disbelief as I managed to hoick the escaping nappy back onto Mabel's bottom, shoving it up her legs and wrapping it firmly around her woolly tummy.

'Honestly pet, I don't think I'll live it down if people see a lamb in a nappy walking down the road on a lead.' Steve's eyes narrowed. 'And can you imagine what they're going to say when they see me in the mart?'

'Ah, you'll be fine,' I replied, holding up the end of the gaffer tape and ripping a bit off with my teeth. 'It'll be no worse than that mad bugger who goes into Hexham Tesco with his pet duck wearing a bow tie.'

I wrapped the gaffer tape tightly around the droopy nappy and stood back to inspect the result.

Mabel was racing between the dishwasher and the vegetable rack, enjoying the noise of her hooves tippy-tapping on the tiled floor, jumping and twisting, just like a bouncy ball. So far, so good, and the gaffer tape seemed to stop anything falling out of the side of the nappy legs.

As my friend had said, you can't just take a lamb off a farm and transport it across the countryside; you need an 'animal movement licence'. These came into effect after the foot and mouth crisis, and a very good idea they are too in preventing the spread of disease and controlling the movement of farm stock. I rang the local council to explain that I had a pet sheep that was going to a book signing.

The man on the end of the phone seemed completely out of his depth.

'You need a license to move a sheep?' he said at first, sounding unhappy about the whole situation.

'Not just a sheep, a pet lamb. On a harness and lead. And in a nappy. I'll be taking her to different bookshops and then bringing her home again. Although we might have to stop for petrol. And maybe something to eat,' I replied.

'So… the animal won't be moving onto another farm. Or going to slaughter?' he queried. I could imagine him sitting in a busy council office, clicking his biro and hoping someone else would take over.

'No. No one is going to buy her. Or eat her. Mabel will just be going there and back.'

There was a pause at the other end of the phone.

'Well, we do have people who want a circus licence. For a performing animal.'

'But I'm not dragging her around the country to go on stage or anything,' I replied. 'For one thing, I haven't got enough nappies.'

There was a long silence.

'I think you might have to ring the DEFRA vet to make sure this is okay.'

'The DEFRA vet?! What, like the Big Chief vet of the entire UK? Are you sure?'

'Yeeeees,' he said, sounding a little more confident. 'I'll give you the number and you can just check.'

So I rang the Lord High Chief Vet, who terrifyingly was an actual Sir and had lots of letters after his name. He was just the nicest and roared with laughter when I explained about Mabel and the bookshops. And the nappies.

'My assistant will send you an online form to fill in – it's a special temporary animal movement one. You'll need to send it to me every time you visit a new place. Just bing it through on email. Good luck in the bookshops and don't forget sheep love the taste of paper.'

For every bookshop visit I filled in a form and sent it off to this very nice, and very important, vet who promptly signed it and sent it back straight away.

Armed with a large dog crate and a stack of nappies, Mabel and I set off to the first bookshop signing. It was in Ripon, North Yorkshire, and we tootled down the motorway quite happily: Mabel fast asleep on her lamb bed tucked up in the crate and me nervously eating liquorice allsorts whilst keeping an eye on her duct-taped nappy. The first problem was walking Mabel from the car park to the bookshop. Unlike a dog, Mabel wouldn't walk alongside me or slightly in front but preferred to walk directly behind me, following behind in a line. Sheep naturally follow each other in single file, so the tiny lamb obviously felt much more secure tucked in behind me. If I stopped, she bumped the back of my knees or wrapped the lead around my ankles. It was very slow progress as I kept having to disentangle myself or nudge Mabel into the gutter to keep her out of the way of very interested passers-by. The entire population of Ripon seemed to appear when Mabel trit-trotted down the High Street: children and adults kept stopping to pat her head, cars tooted as we walked past and any dogs on the lead kept a wide berth or went absolutely mental – barking and yapping at the strange woolly creature. Mabel wasn't bothered at all; as far as she was concerned she was just out for a walk with her mum, although she'd never seen so many cars in her life.

Once we reached the bookshop the little lamb was a huge hit. I'd brought a bottle of milk and she settled herself down on a rug in the centre of the children's section and gracefully accepted any pats or strokes. After a short nap and nappy change, she grew bored and started to mooch along the bookshelves, carefully tasting the paperback spines and nibbling on the pages. The bookshop had an outside area with grass and picnic tables, and I carried Mabel out and took off her lead so she could belt from one end of the garden to the other, burning off some of her excitable lamb energy. Back inside, Mabel continued to eat the bestsellers and the booksellers started to look rather alarmed, so we decided to call it a day and made the short walk down Ripon High Street back to the car. I lifted a yawning Mabel into her crate, strapped a fresh nappy around her middle and we set off home. Halfway up the A1, I realised I was starving and made the snap decision to pull into a McDonald's at a motorway services. Mabel bleated to be let out of her crate, and as we trundled along in the drive-thru queue, I lifted her out and sat her down in my lap. The lady at the window peered into my car did a double take as she saw the little woolly lamb staring back at her through the glass. Wide-eyed, she handed over the food, and I drove into the car park, where I ate the chicken and Mabel thoughtfully chewed on a lettuce leaf or two. Whilst I was eating, a few McDonald's employees hesitantly approached, and I opened the door so they could see her sitting regally on her cushion on the front seat, sucking on a tomato. I then answered many, many questions about farming and sheep and why I had a 'baby lamb' in the front of my car.

After the first book signing, Mabel and I became more adept at finding bookshops and places to park, and we had fun visiting various coffee shops across Yorkshire, Cumbria and Northumberland.

Lambs grow very fast, and over the next few weeks I bought larger and larger nappies from Tesco until Mabel was wearing huge toddler pants and a long string of parcel tape to keep the whole contraption from sliding off her back end.

'Christ,' said Steve one day, eyeing Mabel and her massive nappy as she swaggered around our back garden, 'you'll be needing men's incontinence pants next.'

Mabel was the perfect lamb to take on bookshop visits as she was calm and happy as long as she was by my side. I think she just accepted that her human mum had to travel to strange places, but as long as she had bottles of milk and the chance to sit on my lap she felt safe and secure. We visited a local care home where she had many cuddles with residents, our local opticians and the village primary school, where she was given a school sports shirt to wear and made a tour between all the different classes, bouncing from side to side across the sports hall.

I took Mabel into the nursery and a gaggle of miniature children in tiny uniforms swarmed across to have a pat and listen to my very short spiel about our sheep and lamb flock. All except one tiny girl, who looked at Mabel in disdain and continued colouring in with deliberate concentration.

'Do you want to come and say hello?' I said.

She looked up and narrowed her eyes, shaking out her blonde curls.

'My dad has five hundred yowes. We've got *loads* of pet lambs and some of them live in our kitchen,' she said with a dismissive shrug. I had a fleeting moment of sympathy for the unknown farmer's wife who was probably knee deep in lambing and bottle-feeding pet lambs.

'Ah, that's Ellie, whose mum and dad farm up near Matfen,' said the hovering nursery assistant.

Ellie turned her page around and started colouring in the other side. 'Yup,' she said, with her tongue poking out of one side of her mouth. 'And *our* pet lambs are *much* bigger.'

What a star. Such is the risk of taking farm animals into a school that includes many farming families within its catchment area.

Mabel is now a mother of her own with two chonky lambs, and she has settled happily into the life of an ordinary sheep. She's still very friendly, and when her lambs are weaned she dashes across for a neck and belly scratch. But when she has her lambs at foot, she's a little more cautious and aloof. A typical grown-up pet lamb, she's completely unafraid of humans or Mavis, our sheepdog, and will go in the opposite direction to the flock, especially if she spots a sheep feed bag. She tends to walk just behind me when we're moving sheep from field to field, unconsciously replicating how she kept close as a lamb, which is very cute but also very irritating as I'm always having to persuade her to stay in the field with the rest of the flock, rather than trit-trot behind me into the house. Mabel just considers herself part of the management and expects special treatment. She's a great mum and spends her days in our back field, snoozing under the trees in the heat of the day and peacefully chewing her cud.

I thought that being a farmer's wife for over fifteen years would make me less emotional about our animals and more accepting of those who died when they were young or unwell. But in fact it's the opposite: I've found myself making deeper connections with our flock and I'm embarrassingly devastated when we lose a young animal.

'Ah,' say the militant vegans, 'but you're going to kill these animals eventually, so why are you bothered?'

True – we do send the sheep to a sale to be bought for meat or sold to another farm. But I'm bothered because I want to give them the best life possible beforehand and treat them with kindness and dignity before they're sent to be humanely slaughtered by a licensed abattoir. But some animals are never destined for the mart and become what Steve calls 'field ornaments', and I'm floored when they die well before their time.

This year, Wonky was born as part of a set of triplets to an extremely grumpy Suffolk ewe. Steve called me from the lambing shed.

'When you've got a moment, can you check pen twelve? There's a small triplet that can't stand up.'

I pulled on my boots and coat. At the height of lambing Steve doesn't have the time or energy to look after the more needy babies, so I've taken on the task of nursing these lambs until they can manage on their own.

We peered over the edge of pen twelve whilst the Suffolk ewe stamped her feet in irritation. She had pushed the pair of bigger lambs into the centre of the pen and was busily licking them clean, but she was ignoring the smallest baby and it was lying half in and out of the pen. As I peered at the tiny family the ewe moved her feet to reach the head of the biggest lamb and unthinkingly stood on the smallest, making it squeak out in pain. I reached down instinctively and picked up the bleating baby.

'Aye, she doesn't want it and has been bashing it around since it was born. It can't stand up,' said Steve.

I cradled the little body whilst I checked it over. It was a ewe lamb and had a jaw that was badly undershot, squashed nostrils and a skew-whiff mouth. Her front legs were oddly long and bent at right angles at the knees, the tiny black hoofs twisting round so that she couldn't get to her feet.

'Do you want to give it a chance? It'll probably die, so I'll put it down if you think it's too bad,' he said.

I cuddled the shivering lamb to my chest and slipped its tiny body down the front of my coat to warm it up.

Steve raised his eyebrows and gave me a quick smile before walking off to the back of the shed, slowly shaking his head. He knows that I give *everything* a chance.

The little lamb squirmed against my jumper, hungry and cold, so I plopped her down in the straw under a warm heat lamp and mixed up a bottle of lamb milk. The lamb hadn't sucked at all from her mother, so I 'tubed' it with some colostrum, gently pushing a flexible tube into its gullet and pouring down some artificial milk into its stomach. Immediately her tiny tummy looked much fuller, and I carefully lay the baby down in the straw to have a proper look at her twisted jaw and legs.

She was a very peculiar sight, with a squashed pug face and a jaw that jutted out at right angles, giving her a look of defiance, like a titchy prize fighter. Her legs were very deformed and were far too long for her body. The extra length meant that her shoulders and knees bent outwards, making it almost impossible for her to walk without collapsing at the shoulders. And her feet were just as strange. Instead of two straight, cleated hooves that pressed flush against the floor, the separate hoof sections were at ninety degrees, contorted so that they stuck out sideways, meaning that she was walking on the inside of her foot. Her back legs weren't so bad and the rest of her body looked normal, although I obviously couldn't see under the surface. God knows what strange organ or skeletal configurations were covered up inside her woolly body.

'I hereby name you Wonky,' I said, as I lifted her onto my knee, scratching under her tiny chin.

She wriggled in my hands, enthusiastically nudging my fingers, instinctively trying to look for milk.

Even though she was strangely built, Wonky was full of life and didn't seem to be in any pain. Once she got the hang of the bottle she bleated loudly and thumped it with her head to try to make the milk run faster into her mouth. She couldn't walk at first, so I started some basic physio by lifting her under her tummy with my welly boot and holding her chest so that she could get the hang of how to lift herself upright. The long front legs gave her the look of a tiny woolly crab, and her back had a sharp slope from her shoulders to her tail.

Every four hours I'd go and sit in the straw, holding a warm bottle as Wonky bleated in welcome and tried to wiggle her uncooperative body to sit on my knee. Her favourite position was to sit over my left leg, her hind legs on the floor and her front legs tucked under her chest as she rested on my knee. After drinking her milk in hungry gulps, she'd sigh heavily and relax into my arms, settling down for a snooze as I scratched around her ears and the bud of her horns.

Wonky was *so* appealing. She was a tiny fighter born into a strange, distorted body, and she was doing the best that she could with what she had. The fact that she sighed and snuggled into my lap was so cute, and her happy bleats and squeaks when she saw me (and the bottle) made my heart melt.

She gradually grew stronger and, encouraged by the rest of the pet lambs, developed a way to stand and walk – hauling herself to her feet and using her front legs like stilts. I massaged her tendons every day and looped bandages around her chest and shoulders to try to anchor her joints into place.

Wonky was born in the midst of the first Covid lockdown and the vet spoke to us over the phone, advising painkillers and a splint for her front legs. And gave me a big dose of reality.

'You know she's probably not going to make it, Sal,' said Dave the Scottish vet.

'Aye, but I just want to give her every chance.'

He sighed and I could imagine him wandering around the surgery, the phone wedged between his chin and shoulder. 'I'll try and get some splints sorted. But if she doesn't walk and she's in pain, then well… you know.'

'I know.'

Wonky was immensely popular on Twitter, and soon I had a little coterie of Wonky followers who called themselves #TeamWonky and oohed and aahed over her photos and videos. Lockdown was so boring and stressful that in retrospect those social media posts must have provided a little lamb-shaped ray of sunshine in people's lives, as well as furnishing me with lots of suggestions and ideas to try to straighten and strengthen her legs.

One day a friend of a friend (who will remain anonymous) turned up late in the evening decked out in full scrubs, mask, apron and gloves. She had just finished a long shift as a hospital nurse, but after seeing Wonky's photos on Facebook she had taken the time to collect some plaster casts and bandages and come to our farm to try to help.

It was dark outside, so I picked up a torch and guided the nurse through the farmyard and into the warm light of the pet lamb pen, and she knelt down carefully in the straw to inspect Wonky's strange front legs. After a bit of thought, she picked up the smallest coil of plaster bandages and carefully wrapped them around Wonky's bent front feet, creating two rigid, plastered legs that stuck out straight in front. Wonks was mightily confused and sat on the pen floor, her front legs splayed out in front, bewildered and refusing to stand up. When the casts were set and the nurse had dispensed advice on how to keep her casts clean and dry, she walked back to her car, ignoring any suggestions of payment, cups of tea or public votes of thanks on Facebook. God bless her, and the NHS.

Wonky was nonplussed by her new stiff legs and for a while, struggled to get onto her feet. Because her front legs were too long for her body she couldn't get them under her chest, and when she tried to twist herself into a better position, her body rocked slowly over onto its back, with her cast legs sticking straight up into the air. With a sharp pair of shears, I cut down the bandages so Wonky could move her shoulders and that eventually made it possible for her to lurch to her feet and wobble unsteadily over to the milk machine.

'Clever girl,' I murmured as Wonky, now with a sloshing tummy full of milk, teetered over the straw on her stilt-like legs to sit down with a thump under the heat lamp. I left her there, her eyes drooping as she began to make tiny, thrumming snores.

The next morning, the casts were still on but covered in lamb poo. Wonks had obviously had a great time hobbling around her pen with her flock mates, trying very hard to keep up with their high jinks. I had to check her front feet every day for soreness or swelling but all seemed to be well.

Wonky could walk, after a fashion, and the casts provided her with a bit of extra support, but her misshapen jaw was another matter: her squashed nose meant that her breathing was always a bit laboured, and the jutting jaw made it difficult for her to chew. I grew used to seeing her nibbling on pieces of hay and chewing her cud – her teeth were always visible at one side of her mouth, so it looked like she was giving everyone a lopsided smile as she concentrated on her dinner.

Wonks became a superstar, taking trips with me into the field outside, meeting visitors at the brewery and generally living the life of a well-pampered celebrity. Our friendly nurse came back to check on her several times, and eventually we took off the rigid bandages. Wonky frequently looked sore on her distorted front feet, and she spent a lot of time kneeling down, but with painkillers and the occasional sock-like bandage she seemed to be comfortable enough.

As she spent more time outside her deformed toes started to create pressure sores on the inside of her front feet, and I tried to find a way to relieve the pain and soreness. At first I experimented with a pair of 'shoes' made out of padding and sticky vet bandages, but they only stayed on for a couple of hours, and I frequently found them either in next door's pen or trampled into the straw bedding. At this point I was in touch with a farm charity in Wales and, on their advice, I bought some cheap dog's paw protectors, which combined with lots of antiseptic-drenched cotton wool seemed to keep Wonky's feet clean and pain free. A friend who had retired from farming came to visit little Wonky and looked on in silence as I proudly showed him Wonky's new shoes and how I was dealing with her problems.

'You're spending all this time on the little thing, but it'll never get better. What are you going to do with her?'

'She'll stay with us forever,' I said, defiantly raising my eyebrows at Steve who was resting his arms on Wonky's pen.

'She will bloody not,' said Steve. 'I should have put her down when she was born. All this effort and for what?'

Our friend nodded his head and cleared his throat. 'Well, it keeps her occupied,' he said, thrusting his chin to show he was talking about me, 'and god, my wife was exactly the same.'

Wonky loved being outside in the paddock with the rest of the pet lambs. Her particular friend was a tiny ewe lamb called Fern, and they trundled around the field together, keeping themselves carefully separate from the rest of the flock, eating grass and snoozing in the sunshine. Wonky would bleat in welcome when she saw me open the gate and would hobble over to nudge my knees, trying to get me to sit on the grass. When I did, she would launch herself into my lap, her front legs tucked underneath whilst I stroked her head and whispered sweet nothings into her large, bat-like ears.

She wasn't walking any better and the hooves on her hind legs were beginning to curl into weird shapes, forcing her to walk on her tiptoes. It was also a problem when we tried to move the pet lambs into the sheep pens to worm them or give them their vaccinations: Wonky couldn't walk fast enough and her feet were too sore on the cobbled yard.

I'd been asked to speak to a journalist for a national farming magazine, and after droning on about our farm and my first book, they sent me an email to ask whether I'd be happy to pose for a photographer. I've had various photos taken of me and the family over the years, usually with us parked in front of some haybales, gurning into the camera or posing with some bored farm animal that we'd bribed to stand still for two seconds.

The photographer snapped off a few pictures of me posing in front of straw bales and the tractor and then asked if I could pose next to a couple of sheep. I decided Wonky would be perfect, and along with her friend Fern, I crouched down beside them as the photographer made squeaky noises so that Wonks would look into the camera.

The first time I saw the photographs was in the pages of the glossy magazine, and I stared in admiration, admiring Wonky's thrust-out chin, her involuntary smile and her bendy front legs. Steve was less pleased.

'Couldn't you have picked a sheep that looked normal?' he said, peering at the photo from over my head.

'She's beautiful in her own way,' I said, primly shaking out the pages, 'and anyway it's a talking point as she looks so unusual.'

I shared the article on Twitter and the reaction was mostly positive, although a few did query about whether Wonks was in pain or tentatively gave the opinion that she should have been put down when she was just a newborn lamb.

Wonky reached six months and spent her days snoozing in the paddock, hobbling around eating grass or wedging herself under the lamb feeder when it was raining. During the summer I noticed that the little lamb had stopped growing and Fern and the rest of her flock mates started to outpace her in almost every way. Wonks then started to lose weight and to appear tired and lethargic, uninterested in any lamb feed and spending a lot of time lying asleep beside a contentedly chewing Fern. I knew that the end was nigh, and it would soon be time to make the difficult decision.

The day came sooner than expected. We had herded the pet lamb flock into the sheep pens for a worming drench, and I had to carry Wonky as she couldn't keep up with the bigger lambs. Steve carefully syringed the medicine into Wonky's mouth and started to 'dag' her poo-stained bottom, clipping off the soiled wool around her backside. Wonky was so small that she didn't fit properly into the sheep race, and she slid through the grippers that normally held a lamb safe and secure. Suddenly Steve swore and jerked back.

'Bollocks. She moved and I've caught her leg.'

I looked over the side of the race. Wonks was holding up a back leg in pain, with a line of blood across her hock.

Steve lifted Wonky out of the race, and she sat on his knee as I investigated the cut.

'You've only nicked her. It's not deep. I'll spray some stuff on it and carry her back to the field,' I said.

We kept an eye on her that evening and she refused to get to her feet, preferring to lie on the grass with her legs tucked underneath. I lifted her to the water trough and held her up under her chest so she could take a drink. The rest of the pet lambs looked absolutely normal, with her friend Fern grazing beside her unconcernedly. Wonky just wanted to sit on my lap, so I sat down on the wet grass and lifted her onto my knee, petting her head and ignoring the damp that soaked through my jeans.

I knew it was time. She was in a spot easily seen by the glampers, so I carefully carried her over the brow of the hill to a more private area and rang Steve on my mobile.

'Can you come and put Wonky down?' I said, sobs rising in my throat. 'She can't walk and she's in pain.'

'Aw pet, I'm sorry. I'll be there in a short while.' He was in the pig shed, so I sat and waited on the grass with Wonky in my lap and cried into her wool, apologising through the tears and rubbing around the base of her horns and under her chin.

Wonky was quiet and seemed happy just to sit cuddled up on my knee, and she only raised her head when Steve drove up on his quad, clutching the humane bolt gun on his lap.

I kissed Wonks and put her gently on the grass as she bleated in protest. I stood behind the quad, head turned away as Steve quickly knelt beside the tiny sheep, fitting the bolt gun to the centre of her forehead. I couldn't believe how much I was crying – my nose running, tears pouring down my face as I heard the 'click!' of the gun behind me.

'It's done, pet.'

Steve wrapped me in his arms, and I sobbed hopelessly into his jumper.

'I don't know why I'm so upset,' I said, my breath hitching in my throat. 'I've had lots of pet lambs that died. But she was so special.'

Steve didn't say anything but hugged me until my sobs turned to hiccups. He gently collected the little body and took it into the shed as I hugged myself and, eyes swollen with tears, went to ring my mum and cry some more down the phone.

Deep down, I knew that Wonky wouldn't survive to adulthood, but she had such a passion to live and an appetite for milk that I couldn't bring myself to end her life as soon as she was born. Wonky had a very small and a very short life, but she managed to touch a lot of people through my posts and photos on Twitter, Facebook and Instagram. I feel like I'm justifying the time I spent on Wonky, but surely every animal deserves a chance no matter how small or insignificant they might be.

*

Lambing is hard work, and it's even harder when you have to do it surrounded by members of the public whilst trying to demonstrate good working practice and trying not to have a meltdown. Steve and I have had some arguments, usually while moving sheep, right in front of a crowd of curious onlookers on their way for a cuppa in the brewery tearoom. A couple of years ago a television company wanted to come and film us for a news segment on lambing, and after being up most of the night, we met the presenter and cameraman at the ridiculous hour of half past four in the morning as they wanted to film the sunrise over the farm. Steve and I spent the rest of the day gently bickering, unravelling slowly with tiredness. The producer used a drone to film us from above herding some pregnant ewes through the paddock gate and into the shed. When I saw the finished film, I was pleased to see that the drone had been too far away to capture the steaming argument we were having during the filming: all about the sheep and why I couldn't follow instructions and how Steve expected me to help when he wouldn't give any clear idea of where he wanted the ewes to go.

Each lambing season brings another carousel of new lambs and their personalities. After June, the lambs are big enough to be turned out at grass and I leave my lambing assistant duties behind, gradually forgetting about the long days and hard work until I'm rudely reminded once again when March and April swing around.

Aside from the pure busyness of lambing, our flock needs care and attention during the rest of the year: from worming to feet trimming; dagging and shearing; vaccinations and veterinary care. It's a packed timetable. The heavy pressure of farm work and animal care means that sometimes family life becomes less of a priority, and life deteriorates into one long farm to-do list. I try to make a distinct effort to carve out spare time with both Lucy and Ben, especially during the school holidays, and organise some days completely away from the farm. The children are very good at entertaining themselves, but I feel very guilty if they're left to their own devices for hours at a time. Now that they're both older, they can come and help us outside, but I also try to spend time with them on non-farm-related activities, as I never want them to feel that they are missing out.

However, trying to keep a balancing act between home and farm life sometimes means that we rush through tasks in an attempt to get them finished – with disastrous results, as we found out in September 2021.

We were planning to foot trim the tups in preparation for their assignation with our in-season ewes, and Steve was on top form, ebullient and in charge, as he hurried each ram through the sheep race and into the handling system. We planned to finish the task in one afternoon, before the kids came home from school at four o'clock. Our tups are massive, solid squares of muscle and their shortened Texel faces and truncated nostrils often lead to breathing problems and wheeziness; it was imperative that they were checked in the next couple of days as tupping season was looming and we were already behind. It's almost impossible to hold an adult tup still to check them over and trim their feet without using specialist sheep-handling equipment, and we had just invested in a new sheep turnover crate, so we could tip them onto their sides to clean up their hooves and check that all their bits were pain and lump free.

Our first tup in the stocks was Thrusty Clappernuts, and he lay on his side, eyes rolling in frustration, as Steve checked his teeth, looked in his ears and palpated his testicles, checking for any tell-tale bulges or swellings. All seemed well so Steve picked up his extremely sharp foot-trimming shears in preparation for snipping and peeling away any excess hoof, just like we do to our own fingernails. As soon as Thrusty felt his leg being touched he snorted and kicked out wildly, bashing Steve's arm so he drove the shears violently into his other palm, almost obliterating his entire left hand and spraying an arc of blood and bone into the air.

'Oh dear,' said Steve, swaying slightly as he watched the blood gushing from the gaping wound. He tried to put out an arm to stop himself falling over, but the sudden pain made his knees buckle and he fell face forward onto the filthy floor.

'Oh shit!' I shouted and lurched forward to try to pick him up. Eventually I got him into a sitting position and inspected the damage.

His middle and index fingers had been almost sheared through, the bone splintered and the skin and tissue ripped open so that his fingers were attached to his palm by a shard of bone and a sliver of gristle.

Thrusty was still lying in a prone position, watching the goings-on with curiosity and with maybe a tinge of triumph, so I yanked the controls to turn him upright and release him back to his mates.

Blood was still streaming down Steve's arm, staining his overalls and puddling on the floor.

'Oh god!' he groaned. 'I think I've buggered it up this time.'

Farming is very, very dangerous; I've lost count of the times that Steve has cut his fingers or dropped a heavy piece of equipment on his foot or bashed open his head. He does his best to minimise risk – we can't afford health or life insurance – but he's a frequent flyer at our local A&E where the nurses greet him with weary amusement every time he hobbles into the waiting room, leaving a trail of blood on their nice, clean floors.

I wrapped an old jumper around his hand, hauled him into the farm truck and drove him straight to A&E. He was taken almost immediately into surgery. After a few frantic phone calls Mum rushed back to the farm in time to meet Lucy and Ben from the school bus and to give them their tea. A few hours later Steve emerged, hand swathed in a massive bandage, being supported under one arm by a tiny nurse.

'He fainted on the table,' she said, lowering him gently into a seat. Steve was as pale as a ghost, hand cradled in one arm, his eyes wet and bloodshot.

The surgeons had managed to save his fingers, stitching the tendons back together and carefully reconstructing a flap of skin to repair the deep gashes made to the muscles. The bones would take a little longer to fix. Steve looked groggily at the huge bandage around his left hand.

'You'll need to keep that hand elevated,' said the nurse, handing over a big box of antibiotics and painkillers. 'It's going to be sore at first, so lots of rest, and no going straight back onto the tractor and finishing off the jobs. I know what you farmers are like.'

I drove Steve home in silence, his body slumped in the passenger seat as his eyes drooped due to the after-effects of shock and pain. I helped him in, answering hundreds of questions from Lucy and Ben as they helped him up the stairs and into bed. He slept for twenty-four hours, and I made frantic phone calls around my friends and family, trying to find people to help me with the animals. At that point we were also looking after one thousand pigs, and there was just no one who was available to help apart from me and my mum and dad.

Even in a crisis, the farm never stops. No matter what happens, the animals are still to be fed and checked twice a day, day in, day out, all year round. In times of urgency the only people I can ask to help are my mum and dad, and bless them, they have risen to every occasion. Dad is eighty years old this year and Mum is seventy-six, and they take everything in their stride, donning waterproofs and boots to muck out pigs or patiently learning how to drive the quad bike (although Dad still has trouble getting it into reverse) or the best way to lamb a ewe. They do all this with a sense of adventure, with a shout of 'At least we're useful!' as they learn how to catch a ewe or clear a stream. My parents have always had a sense of 'just getting on with things' and 'not letting age slow them down', and they laugh uproariously at their adventures in pig or sheep husbandry. Mum has to wear calipers due to the fact she has Ehlers-Danlos syndrome, but she never lets it slow her down and is constantly on the hunt for a pair of boots that will support her wobbly right ankle but also allow her to trudge through slippery pig muck. Dad is deaf and has a bad back, but he loves being involved and asks endlessly if we need any extra help. Dad is also a great travelling companion and loves accompanying Steve on investigative trips to look at new pieces of machinery or travelling across the country listening to important talks on pig husbandry. They are both full of energy, resilience and good humour, and we couldn't manage without them.

Steve recovered slowly, but it was a long time before he could use his left hand properly again. My mum and dad swung into action and appeared every morning at seven o'clock to help me muck out our pigs. I will never forget watching her – in smart trousers and jumper, her usual necklace in place – up to her ankles in slurry, digging doggedly into pig muck, whilst my dad – wielding a shovel and wearing a terrible pair of overalls – hoiked massive clumps of stinking straw over the pen walls. Whilst Steve completely recovered the use of his hand, the accident really brought home how vulnerable the farm was to illness or injury and how much I needed to rely on other people if anything went wrong.

CHAPTER SIX

....

It Takes a Village

....

The ultimate goal of farming is not the growing of crops, but the cultivation and perfection of human beings.

Masanobu Fukuoka,
The One-Straw Revolution (1978)

Steve and I don't have an extensive network of farming family, and without any paid help we had to rely on the charity and good will of relatives, who even though they're often short on experience are brim full of enthusiasm.

Like most farmers, getting pregnant and having children meant I had no maternity leave or time off and I worked right up to the birth and carried on almost immediately afterwards, trying to fit jobs around constant breastfeeding and naptimes.

It was a challenge working out how to keep my newborn babies safe when I was at work, so at first I plonked them into a papoose, strapping them safely to my chest as I mucked out animals or worked in the brewery office. When they both grew too heavy, I had to find some more creative child-sitting arrangements.

Sitting in the office whilst carrying a baby was a non-productive nightmare, and the office wasn't big enough for a play pen. I also needed something to keep them away from the steep tearoom stairs and from getting into something they shouldn't, such as the cleaning solution or the big bags of hops. Once I spotted a huge cardboard box full to the brim with malted barley and after emptying it all out, Lucy was stuffed inside the rather dusty interior and spent the morning sucking placidly on the cardboard, playing with a toy tractor and waving at anyone who passed by.

Trying to keep the kids safe when I checked on the sheep was another problem. I didn't feel comfortable holding them on the quad bike as I had anxious visions of losing my grip and squashing them under a wheel. It was impossible to drive the tractor whilst they sat on my knee, and the ground was too rough for a pram or buggy. I had the brilliant idea of pressing Candy – our small, fat Shetland pony – into service.

On a freezing cold winter's day when Lucy was just over twelve months old, I dragged a suspicious Candy out of her field and into her stable. After a quick brush, I strapped on her cub saddle, attempting to cinch the girth around her bulgy middle as she pinned back her ears and puffed out her belly in annoyance. Candy hadn't been ridden for a good year or so and had spent an entire summer and autumn in her field, so she was rather barrel-like with a squidgy layer of tummy fat. Lucy, dressed in a red, all-in-one fleece suit and riding hat, toddled with interest around Candy's legs and under her belly whilst I heaved the girth up to the very first hole, ignoring the pony's withering side-eye. I carefully led Candy outside and lifted Lucy onto her broad back, adjusting her stirrups and surreptitiously tightening the girth again whilst Candy snuffled among the leaves on the ground, looking for a tasty snack. Lucy was very excited, and with a bit of biscuit in one hand and the other clutching the strap on the front of the saddle, babbled happily as she lifted her face to the watery sunshine. Candy was less excited and sighed hugely as she felt the tug on the lead rope, reluctantly ambling behind me through the gate into the front field. We marched along companionably, skirting around the outside of the field, with Candy flicking her ears back and forth as she listened to Lucy's chatting, and I scanned the flock, looking for any sheep on their back or signs of lameness and scour. We were doing so well, and I congratulated myself on my idea of child-friendly farm transport until Candy suddenly spotted a pheasant by the stream and promptly pricked up her ears, snorted and planted her feet into the deep turf, absolutely refusing to go any further.

'It's a pheasant!' I said, as Lucy giggled with delight. 'You're not telling me you're frightened of a pheasant?!'

Candy snorted like a small dragon and quickly whipped around to stand behind me, trying to convince me that the pheasant was the most threatening thing she'd ever seen. (NB: She was lying. She had seen plenty of pheasants in her own field, and I'd even watched her fall asleep as one pecked around her feet.)

'Oh my god,' I said, pulling on the lead rope so that she skittered along behind me. 'Stop pretending. We are definitely walking around the rest of this field.'

She gave one last huff of pretend fright and started off again, making it very obvious that she thought being expected to check

all the walking and exercise. When Ben was born, I continued the tradition, plonking him on top of Candy whilst Lucy sat on Nemo, our bigger black and white cob. I still have a fantastic photo of Ben sitting on a very tubby and cross-looking Candy with Lucy grinning next to him, on the back of a doubtful-looking Nemo.

As the children grew up, they toddled around after me, helping to muck out or feed a lamb. We had and still have a very strict rule that children have to keep out of the machinery sheds as they're full of dangerous equipment. I'm always terrified of kids around tractors and telehandlers as the poor visibility makes it very difficult to see properly when you're in reverse. Horrifyingly, it would be so easy to accidentally squash a child. After many lectures my kids know to stay well out of the way of any farm vehicles, and they know by experience how dangerous a farm steading can be. One morning, when Ben was three years old, I was tinkering in the machine shed. Ben was playing on his bright green push-along tractor and was zipping around the farmyard making happy 'brrmmm, brmmm' noises. I took my eye off him for a second, and he pushed his way into the machinery shed to stare at the big machines. He scooted along until he was near the huge round baler, and when his little tractor got stuck in the corner, he tried to step off to turn it around. His trouser leg caught on the steering wheel, and he fell forward, gashing his head on an upright prong that stuck straight up on the back of the baler.

There was blood *everywhere*. I nearly passed out from the sight of a screaming toddler with blood gushing from a deep cut in his forehead. After a trip to the local A&E, some stitches and a lot of Chocolate Buttons Ben came home cuddled in my arms, woozy from the painkillers and his eyes swollen and red from crying. That was the last time I took my eyes off my kids around the farm. There is so much that could go wrong, and it was only thanks to luck that Ben escaped having a much more serious accident. As it is, he now has a nifty Harry Potter-esque forehead scar.

Lucy and Ben are now teenagers and whilst they do help us, we don't ask or expect them to do more than the occasional chore around the farm. I'm not a great believer in making children work in a family business, especially if they have their own tidying and cleaning responsibilities at home, never mind the piles of homework that need to be done every evening and weekend. Lucy can drive the telehandler and enjoys spending time with her dad, learning about the farm and the animals. Ben, in comparison, tells me that he wants to live in a place where there's lots of people and neighbours as he's very sociable and gets bored playing on his own. He watches his friends in envy when they talk about popping round to a friend's house or going to the local park together. We have to drive to see his friends in the local village, and he grumbles about having nothing to do at the weekends. 'But you have a whole farm to play in!' I say. But I forget living on a farm is not some magical lifestyle to him, as he's been here since he was born and it's natural that sometimes he would like to share his games with someone else.

'Oh, they must have an amazing lifestyle. I'd love to bring up children in the countryside' is a refrain we frequently hear from our visitors. But my children have the same pressures as those in towns and villages, and better scenery and a great view don't always make up for the fact that the local transport is crap and they can't pop down to the shops or visit their friends under their own steam. I hope that we can give both our kids normal childhoods and they find their own paths and interests to follow. Lucy and Ben are now in high school: obsessed by the Xbox and TikTok, the latest online influencers, the latest fashion and music artists and sometimes keen on the farm, sometimes not. Just how it should be.

*

As we can't afford to actually pay for someone to help out, I normally persuade my teenage nephew Matt to make the journey up to Northumberland to give us a hand over the lambing season. As the pandemic hit in March 2020 and we went into lockdown, I became very anxious; firstly because we were about to go into lambing and we couldn't expect any help from Matt and the usual team of visitors, but also due to the even more nightmarish scenario of Steve catching Covid and becoming so unwell he couldn't manage the farm. Later on, the vaccination roll-out, at that time, was concentrating on the very elderly and infirm, so Steve, as a (fairly) healthy forty-seven-year-old was a long way down the list. Across the UK, farmers and smallholders were asking the same worried question: what would they do if they caught the virus? We had no one nearby who could help, and we couldn't expect friends to come and help me with the animals or machinery, as the risk of catching Covid would be too great. If Steve did become ill there was literally no one experienced enough who could step in, and I couldn't manage the heavy work on my own. I tried to cope with the anxiety by avoiding the news and concentrating on each day as it came, but I frequently found myself awake at four o'clock in the morning, ruminating and obsessively refreshing Twitter in case I missed an update. My brain is not designed to cope with huge lumps of stress, and under pressure I tend to flail around with anxiety and, not content with just worrying about big stuff such as a pandemic, my mind likes to ramp up the pressure by fixating on other random worries. I realised something had to change when I found myself at two o'clock on a dark Thursday morning fixedly watching YouTube videos on how to fly a 747 in the event of the pilot being sucked out of the window.

Steve and I decided that Mum and Dad should leave their comfy bungalow and come and live with us for however long the strict lockdown regulations lasted. They were game, bless them, setting up camp in our spare room and throwing themselves into helping us farm: my eighty-year-old dad accompanying me on my evening lambing rounds and Mum constantly in the kitchen, providing stability for Lucy and Ben and making enormous dinners for us all. They were beacons of good humour and resilience, and their ability to roll up their sleeves and just muck in was a huge bonus and help. It was good having them with us; even though we were living through a terrible crisis, they helped by providing a sense of normality and told us about their own parents' struggles during the last war, and how they managed when everything seemed to fall apart.

Every day brought new tallies of those who had died, and Dad kept a diary, carefully noting down in pencil how many of those deaths had occurred in our area. Mum started knitting matching woolly hats for us all, using scratchy acrylic wool she found at the back of her cupboard, and crafting coloured pompoms out of scraps of felt. We fell into a comforting routine as the lockdown dragged on, pooling our resources to make the occasional trip to the supermarket, keeping an eye on any elderly neighbours and trying to home school both children, with varying results.

The farm was weirdly quiet. The brewery and tearoom closed, and the glamping site shut down. We were used to a busy farm-yard with lots of tourists and parked cars, but it now lay quiet and empty, apart from the whoops and shouts of the children: they took advantage of the sudden emptiness, riding round and round a rickety bike assault course that they'd made with old bits of wood and broken sheep pens. Closing the brewery, tearoom and glamping site brought extra stress as we worried about how to pay the mortgage on the buildings and, alongside many other business owners, wondered how we were supposed to manage with no income throughout what would be the busiest time of the year. As lambing started, I became envious of those who were furloughed and posted glorious Instagram pictures of themselves lazing in their gardens or enjoying a sunny barbecue. In comparison, Steve and I just had to make do, and we slogged through each day, trying to share out the sheep work between us and my parents so that we all had at least some time off to catch up on sleep.

As is his nature, my dad looked on the bright side and told us that he thoroughly enjoyed himself staying on the farm. He bought a pair of green, Flexothane dungarees, and every night, the fabric swishing noisily, he would march out purposefully into the dark, waving his torch around, ready for our 5 p.m. to 9 p.m. lambing shift. For an eighty-year-old he was full of energy, and his unfailing enthusiasm and common sense meant I enjoyed our evening rounds as we sat on haybales in the lambing shed, waiting for a ewe to give birth and chatting about the farm, his life and everything in between.

We called ourselves the 'Z Team' as neither of us was very experienced, or very agile. Dad was very deaf and had splashed out on a very expensive pair of hearing aids; he was constantly paranoid that they'd fall out of his ears and he'd lose one or both in the thick layer of straw at the bottom of the lambing shed. At first, he just didn't wear them at all, but it became a bit of an issue as he couldn't hear a thing.

'Dad, can you get me the baler twine?' I'd say, trying to tie up some sheep hurdles before anyone escaped.

'What?'

'The baler twine!'

Dad stared at me from the middle passageway with his hands cupped behind both his ears. 'What?' he'd repeat.

'The BALER TWINE!' I'd bellow across the shed. 'Oh for god's sake, I'll get it myself.'

We decided that his poor hearing was definitely a health and safety issue because if my dad couldn't hear us shout, he certainly couldn't hear the 'beep, beep' of a reversing tractor, and if he couldn't hear the engine he'd be at risk of being squashed. So we made him wear his very expensive hearing aids but tied them onto a bit of orange baler twine that he slung around his neck. It wasn't exactly a fashion statement, but at least if they fell out they'd just hang droopily around his coat collar rather than disappearing into the straw.

One dark evening in April, Dad and I were sitting on our usual camping chairs in the corner of the lambing shed. We were clutching cups of tea and watching a labouring ewe in the big pen who was marked up as carrying triplets and had been puttering grumpily in circles for the last half an hour. Her water bag had already appeared and burst, but we couldn't see any lamb feet poking out, which probably meant that she'd need help to give birth – maybe the lambs were mixed up inside.

'We'll have to go and catch her,' I whispered to Dad.

He stretched back in his chair, arms above his head and his wellies lifting off the floor. 'Let's finish our cup of tea first,' he said, 'and then see how she's getting on.'

Dad and I were always reluctant to interfere in a birth, not only because it increased the risk of infection but mainly because we were so crap at catching the sheep due to my lack of height and spindly arms and Dad's general age-related lack of manoeuvrability. Steve makes it look easy, stepping smoothly up behind a ewe, grabbing the animal under its jaw and twisting it onto its back in one smooth movement. I'd never managed to pick up the knack and we both found it almost impossible to run fast enough to chase a sheep down through the thick straw.

At last, we drained our plastic cups of tea and shuffled off to pick up the supplies: the crook, lambing gel, lambing ropes and elbow-length gloves. Our shed has big metal barriers between each side of the pen and Dad creakily climbed up one side and down the other, just managing to heave his legs over the top.

I slipped through the metal bars carrying my special catching crook in one hand.

The crook is an essential part of my lambing kit. Steve bought it for Valentine's Day many years ago after he watched me try and fail to catch an enormous Texel ewe whilst being dragged around the shed on its back like a rodeo rider. The crook was made of thick plastic with one curved end and a nifty hook at the other, which was designed to nip around a sheep's back foot so that the plastic hinge snapped closed. I was then supposed to hold on as I was towed around the shed behind 150 kilos of angry, pissed-off sheep.

'Right. You go that way,' I said, gesturing around the back of the silage bale that sat in the middle of the shed, 'and I'll creep up on her from behind.'

The ewe in question was lying down in the far corner and narrowed her eyes at us when she saw us shuffling closer.

I carefully crept within a few feet and glanced around to see Dad just behind me, crouching low on the other side of the shed.

'Right! Grab her!' I bellowed suddenly, and I violently swung the crook towards the ewe's back legs. She shot upright and made a break towards Dad, amniotic fluid spattering the straw behind her.

'Grab her, Dad! Catch her by the neck!'

Dad threw himself bodily across the shed and landed heavily on the ewe's back, whilst I ran towards her, trying to grab her fleece. Dad and I accidentally bounced off each other, crashing into the pen wall whilst I rammed my shoulders into the concrete.

We'd successfully caught the ewe, but Dad was in a very peculiar position, lay over her neck with his head buried in the straw and his Flexothane backside sticking up into the air.

'Are you okay?' I said breathlessly.

I'd managed to scrape my left arm down the rough wall, and it welled with blood from a pair of nasty-looking scratches.

'Oh my god,' said Dad, panting heavily. 'Just give me a minute.'

He managed to swing his legs slowly around underneath him, and keeping his weight on the ewe's neck sat up a bit straighter, wheezing with the effort.

I got on with sliding my hand inside the sheep, checking to see the state of the unborn lambs. They were completely mixed up, with three legs up at the front and a confusion of heads and bodies further back. Putting my finger inside a lamb's mouth I was rewarded by a warm flick of its tongue, so at least I knew they were still alive. I concentrated on carefully sorting out the legs and positioning the first lamb through the cervix, pulling on the forelegs so that it slid out smoothly and plopped onto the straw, heaving its first breath while shaking its head to clear its nose of gunk. The ewe started to make deep mothering bleats and chuckles, nosing the little body as I brought the tiny newborn to lie under her head.

'You can let go now,' I said to Dad, and he sat back gingerly, rubbing the side of his skull.

'I think I've bashed myself on the wall,' he said as he carefully sat down in the straw, his legs outstretched in front of him. He looked alright, if a little dazed, so I felt back inside the ewe and quickly pulled out the two remaining lambs. After persuading the new little family into a smaller pen, I sat down beside my dad and started to compare injuries.

'Are you bleeding?' I said.

'Nope, I think my woolly hat stopped the worst of it.' His hearing aids hung off at an angle, swinging gently at the end of their baler twine.

We clambered stiffly upright, picking up the crook and lube bottle, and climbed the pen gate, helping each other over the topmost bar.

After fixing his hearing aids, another cup of tea and a couple of plasters, Dad was laughing and joking as normal and started messaging his mates over WhatsApp about how he'd singlehandedly hurled himself onto an enormous ewe. The incident did nothing to diminish his enthusiasm for helping out, and the next evening he was raring to go back outside, his dungarees swishing and his baler twine back in place.

'Are you sure you don't mind helping out?' I asked him, pulling on my wellies and closing the door behind me.

He paused to think, as we stood outside in the dark, staring at the glitter of stars above our heads.

'No. I've always liked *doing* stuff and learning new things. I like to be needed. And also, I need to make sure that you're safe on the farm. It's dark outside and you're my daughter and it's important that someone's here to keep you safe.'

I pay him in cups of tea, chocolate biscuits and lots of love. Dad is already planning his lambing shifts for 2022 and occasionally asks me hopefully whether he'll still be needed next year or whether Steve will think he's too old.

'Of *course* we'll still need you!' I reply, and he trundles off, a big smile on his face, his hearing aids firmly wedged into both ears.

In previous years, throughout April we've had a gentle trickle of visitors who came to help us during lambing. Alex, Matt, Lily, Sam and Flynn are my teenage nieces and nephews, and they all made the trip up to High House (often with girlfriends in tow) to experience working with sheep. We set them simple tasks – cleaning out pens and helping to feed and water the ewes. The hot water urn was constantly on, day and night, and there was usually a small gaggle of helpers at the top of the lambing shed, drinking scalding cups of tea and munching on a never-ending supply of custard creams. It was really nice, and compared to the isolation of the rest of the farming year, was a lovely change. It's a chance for us to catch up and share news with the younger members of our family.

One year, during a very busy lambing with many triplets and even the odd quadruplet being born, I had the brilliant idea of contacting our local army barracks, based a couple of miles to the north. The soldiers and their families integrate well into the local community, sending their kids to the local school and joining in our village events. I'd visited the barracks before to drink coffee with soldiers' wives and still experienced a secret thrill when I drove through the gates, mainly because the soldiers on the command post saluted and called me 'ma'am' as they inspected under my bonnet for explosives. Maybe, I thought, some bored squaddies might like a day out on the farm to look at the machinery and animals and do a bit of work? I found the contact details of the regiment and pinged off an email, trying to position the idea of a day working on the farm as an exciting, character-building activity. To my surprise they wrote back and suggested I speak to some officers from the Royal Electrical and Mechanical Engineers as they knew that a group had come back from an exercise and were now all on holiday. Some of the lads were bored and had nothing to do.

The next day an army truck containing three *enormous* soldiers wearing greeny-brown fatigues, red berets and big, black boots turned up in the car park. The soldiers gave me a cheery smile and thanked me for the opportunity to see our farm. Steve had to crane his neck to look at their faces.

'Right,' said Steve, eyeing their impressive height and breadth, 'you can definitely help us in the lambing shed.'

We asked the lads to muck out twenty-five dirty pens, move the filthy straw to the midden, disinfect the concrete and then refill them with clean straw. They set to work... and just kept going, not stopping for a cup of tea or a chat until every single pen was clean and filled with bright, new bedding.

After a short break, all three decided to go and visit the pet lambs, and when I went round half an hour later, I found these big burly men sitting in the pen, each cradling a surprised-looking lamb and whispering sweet nothings into their ears. One soldier had gently tucked a sleepy Wonky into the top of his jacket, so her little head poked out under his stubbly chin. After helping me to bottle feed, they fed and watered the pregnant ewes and gave me a hand moving newborn lambs and their mums out into the fields.

'I've never been on a farm,' said a soldier. 'I didn't realise you were just next door.' He gazed at the brewery thoughtfully. 'This'd be perfect for our next night out,' he said, staring at the line of clean casks standing ready outside the warehouse door. 'I'll see about getting a barrel of real ale into the officer's mess too.'

At the end of the day all three shook my hand, dwarfing my tiny paws in their huge grips, and jumped into the army truck, promising to get back in touch to book the brewery for a 'proper night out'. I was amazed by their willingness to act on orders and their speed of work, but also their compassion and gentleness in handling the tiniest lambs. Wonky and I were a little bit in love with them all. I know their regiment was posted to Afghanistan a few months after the visit and I would imagine they carried out their work there in just the same sort of manner: a mix of determination and enthusiasm plus a dash of careful compassion.

Not all our helpers were so useful. I'd kept in touch with a couple of people from my previous employer in the big accountancy firm in the city. They were desperate to come up to see the farm, give me a hand and see the 'baby lambs'. Steve was beginning to grow sick of the endless stream of people to the farm.

'Stop inviting all and sundry to come and see us. It's the busiest time of our year, and we're just the next tourist destination for everyone. It's like owning a really crappy theme park.'

Still, I'd promised my friends and told them to turn up over the weekend to give us a hand for half a day.

They appeared, in pristine wellies and very tight jeans, and I asked them to help me muck out the pens. Cleaning out the pens properly between each mother and her newborn lambs is essential as it kills any bugs and stops horrible infections such as watery mouth or joint ill. It's the most time-consuming job we have during lambing and doesn't need a great deal of skill, so it's perfect for unexperienced helpers. After giving a quick demonstration, I left them to it and went to start the day's chores, weighing out the feed and washing buckets. After half an hour I went back to the shed, expecting my work friends to have finished at least three or four pens. Instead, the shed was empty with just a half-filled wheelbarrow leaning drunkenly on the cobbles and a fork stuck upright into the straw.

I found all three in the pet lamb pens, sitting in the straw cooing over the babies and inexpertly trying to push the lambs to drink from the automatic milk feeder.

'We're having a cuddle!' said one of my friends in surprise. I closed my eyes briefly and scrubbed a hand down my face.

I had to walk out as I felt my resolve begin to crack, worn down by too much to do and not enough rest. The three girls stayed in the pet lamb pen for another half an hour and then came back to the shed. The wheelbarrow still stood in the middle of the yard, and I knew I'd have all twenty-five pens to do by the end of the day.

'Right. Thanks!' I snapped. And then suddenly I decided to tell them the truth. 'Look, we're not a petting zoo. This is my *job*, and I have to get on with it. You said you would help but you've just sat and played with the lambs. How would you feel if I turned up at your office, flounced around and then went and… and… *cuddled* your marketing department?!'

The girls raised their eyebrows and took a step back. I turned around and stalked off, sniffing hard and already regretting losing my temper. I'm hopeless at confrontation. When I turned around their car had gone and I never heard from them again, although they did post a saccharine post on Instagram with pictures of my lambs with a soppy caption about 'cuddling cute farm babies'. They were very careful not to tag me or the farm in any of the captions.

We weren't always without extra help. For many years, we had a sole farm worker called Bill who helped at the busiest times during the farming year. No one was quite sure how old he was, but he had worked for the Ministry of Agriculture back in the 1960s and 1970s and spent time on farms in the Kielder and Wooler areas as well as at one point being a long-haul lorry driver. He had lost his wife a few years before and now lived in the farm cottage next to ours. It was a mutually beneficial relationship as Bill worked a few hours each week, helping us with harvest and lambing, the stone walling and fencing, and in return he lived in the tiny next-door cottage almost rent free. He was a little chap, tanned and wiry, with a hunched back and wispy feathers of grey hair carefully brushed sideways across his scalp. Years and years of hard physical work had damaged his spine, knee and hip joints, and his hands hung unnaturally low whilst his knees bent outwards, like a crab, when he walked. He wore a pair of faded, blue dungarees tied around the middle with some baler twine, and on special occasions an old suit jacket would be added over the top. Bill smoked at least twenty hand-rolled cigarettes a day, and we could hear where he was on the farm from the loud fits of lung-wrenching coughing, especially if he'd been working hard and had got out of breath.

Bill had a wonderful Northumbrian accent, with the rolling r's and sliding vowels and the local way of saying 'pony' (py-er-ny) and 'water' (wa-tt-er). He used lots of colloquial phrases such as 'cowp ow'er' (turn over) or 'cundy' (field drain). It was a lesson to listen to him talk about his childhood, about how he was told to leave his small village school at fourteen years old and had to go straight to help his dad on their family farm. Bill was a better cattle stocksman than a shepherd, and he hadn't much time for any animal that took patience and delicacy, such as newborn lambs. Nevertheless, he was a hard worker, spending a lot of time mending fences and stone walls, a cigarette poking out of one side of his mouth as he heaved up huge sandstone blocks and nailed fence posts. We knew he missed his wife terribly, but he did his best without her, tending the small vegetable plot in his garden and going to church each Sunday. He liked a drink too, and he used to walk through our farm to the pub on Hadrian's Wall for a few pints each Friday night. The walk back seemed to take a lot longer, and I imagine he enjoyed meandering through our fields and woodland with the occasional sit down for a rest.

Bill boasted that since his wife had died, he did all his own housework and had learned to cook, making huge pots of soup that sat on his hob for days as he threw in whatever had gone out of date in his fridge. We could tell when another pan of soup was on the boil as a pungent, oniony smell would drift through the windows and into our house. I was a bit worried about food poisoning and delicately enquired how long he left the soup out of the fridge.

'Divven't worry, pet,' he replied, 'I boil it up that many times there's nowt alive.'

He also had a thrifty rural habit of picking up anything he found on a walk and throwing it into his soup or stew.

'I'm pretty sure he's scraping up roadkill,' I told Steve at one point. The next day Steve subtly enquired as to the provenance of his latest soup ingredients. Bill nodded enthusiastically.

'Aye, it's a nice bit of pheasant that'd been dinged by a car. I got to it not long after. It's a bit mangled but as long as you pick out any gravel, it tastes okay.'

One day a truly overpowering smell drifted through the window. It was garlicky and onionish but had a pungent, thick base smell.

'What on earth are you cooking?' I asked Bill when I handed him some post over the garden fence. 'It smells very strong.'

Bill looked hurt for a second. 'Well, aye, it was a bit of a fluke like. Me mate picked up a bit of venison a week or so ago, so I've been hanging it in the tool shed for a while. I've just thrown it in the pot today.'

He beckoned me in to have a look. In the dark, dingy living room, stained almost brown from years and years of nicotine, I saw a metal tripod over the fire, with a big-bellied pot hanging below.

'You're cooking it on the fire?!' I squeaked.

'Gives a better flavour,' he said, taking a metal poker from the hearth and giving the pot a quick stir. 'Fancy a taste?'

The pot gave a belch of pure garlic with a bottom note of turnip.

'No thanks. Enjoy your dinner though,' I said as I made a quick exit.

Bill was a great favourite with visitors to the farm, and he could often be found hanging over a gate, rolling a cigarette from a pack of loose tobacco and regaling open-mouthed ramblers and tearoom visitors with the stories from his youth in his rolling Northumbrian accent. In Bill's world, when he was a young lad the winters were always colder, the snow had always been waist deep and his anecdotes were full of him heroically digging out frozen sheep from snowdrifts or battling the 'Ministry' red tape, which always ended with him winning the day and triumphing over some faceless townie bureaucrat. I loved it when Bill started telling stories about the village in the 1950s and 1960s, when children walked five or six miles to school every day and were taught in a two-roomed house. In those days all the tenant farmers were beholden to the elder Sir Blackett who owned (and whose descendant still owns) huge swathes of land around Matfen. In those days it was still a very 'feudal' society, and during General Elections farmers would travel to the village voting station, and once they had filled in their ballot papers they'd be asked on the way out of the village hall whether they'd voted Conservative or Labour, which was carefully noted down in an exercise book. Woe betide those who voted against these staunch Tory landlords. Their name would be mud, and at next rental day they could even find themselves out of a farm. The Blackett family did make many positive additions to the small village, building a brand-new church, the tiny school and a 'reading room' where men could go to read the latest newspaper or periodical. The village also had a rather marvellous 'Temperance Hotel' (the area has always had a strong Methodist congregation) which sat right next to the village pub as a pretty effective visual deterrent against the temptations and evils of strong drink. Visitors to the hotel could expect good conversation and accommodation, but if they wanted a pint they'd have to go next door to the convivial and rowdy Black Bull where they were guaranteed a good welcome. There was a blacksmith and a whitesmith, a village shop, the school, a village hall and a number of horses and carts from local merchants that visited outlying farms selling fruit and veg, or meat or fish or more exotic items, like pins, ribbons and secondhand clothes.

Bill's stories described a slow and simpler type of life and even a golden age of farming, but part of me wonders whether it was as good as he made it sound. I imagine that many must have felt trapped: boys left school at fourteen and had no other options apart from going into the family business, often working on the farm as a low-paid assistant, and the girls had no other prospects apart from going into service or marrying at a young age with no expectation of anything apart from housework and children. The stifling village life must have been hard to escape if you were at all different – Bill never mentioned anyone who was outwardly gay but instead talked about 'confirmed bachelors' with a knowing wink. And everyone knew everything about everyone else: there would be no hiding your personal predilections or transgressions from the local gossips. No wonder many villagers left for the bustle and anonymity of the big Northern cities.

As Bill grew older, we kept an eye on him over the fence, bringing him the occasional bag of shopping or inviting him across for a plate of Sunday dinner. He began to look frailer, with a shrinking frame and an ever-deeper smoker's cough that did nothing to put him off his extensive cigarette habit. He stopped helping on the farm as he couldn't catch his breath when the autumn and winter winds roared around the old farmhouse, and in the summer, the dust and humidity affected him so much he couldn't work outside. His daughter would pop in almost every day, persuading her dad to stay inside during the coldest or windiest weather. Eventually poor Bill was diagnosed with lung cancer, and after months of hospital appointments he moved reluctantly into his daughter's house, fussing as she packed his precious books and photos into cardboard boxes and eventually waving sadly at us from the back of the car as she pulled away with all his worldly possessions piled up in the boot. He told me that he wasn't afraid of dying. 'I'm missing her,' he'd say, referring to his late wife. 'It'd be canny to see her face again.' And it wasn't long before Bill did just that, slipping away in a local hospital with his family around him. I hope he's up there in heaven, arm around his wife, regaling all the assembled multitudes with stories about life in Northumberland and the fights he won against the 'Ministry'.

It's often said that 'raising children takes a village', and I think that refers to a farm as well. High House was never supposed to be worked by just one family, and in its heyday in the nineteenth century would have involved a whole team of staff: from carters to ploughmen, horsemen to shepherds, and their wives and children. Even now with the latest machinery Steve and I can't manage on our own, and we rely on a raft of unpaid family members to help out with everything from childcare to mucking out the pigs. We all depend on each other and trust that in an emergency everyone will down tools and come to help. It's clear to me that underneath all this family support is a strong foundation of love. Love is the emotion that drives my dad to do the late-night lambing shift alongside me for fourteen days in a row, and it is the catalyst to

CHAPTER SEVEN

This Too Will Pass

I think farmers' wives' mental health needs highlighting. (Note: Some in the rural community have engendered a lot of criticism by referring to women as 'farmers' wives'. I don't use this term in a pejorative sense, rather as an umbrella phrase to encompass all those who have a partner who works as a farm owner or tenant.) A recent report on the state of mental health in agriculture makes very grim reading, with a whopping eighty-eight per cent of farmers under the age of forty ranking depression and anxiety as the biggest hidden problems in the industry. Farmers are much more likely than the general population to suffer from depression due to crippling financial pressures, fluctuations in the market, government policies and feelings of isolation. There has been a much-needed upsurge in publicity around agriculture and mental health, but in my opinion it tends to focus solely on the issues that farmers face rather than including their wives and partners. Farmers' wives often have to take the brunt of the stress and anxiety from their husbands, especially as traditionally, male farmers have found it difficult to talk to anyone else about their mental health problems. This leaves their wives to pick up the pieces and keep the family and children together, whilst often holding down their own full-time jobs and careers and dealing with their own worries and anxieties.

A generation ago, women worked hard, but farming was buoyant enough that they had the luxury of staying at home without needing to work for an extra income. There was also a wider extended family, often living very close by, who could pitch in to help look after the children.

Don't get me wrong – it was still tough. Steve's grandmother not only had to feed her husband and three strapping sons but also cook three meals a day for all five farm workers. It must have been relentless, waking up early every single morning to clean and black up the range and keeping all the fires burning in the coldest winters. She also kept chickens, made her own butter and was an amazing baker, but I can't help thinking at least she didn't have to split her focus between her home, her children and working for another employer.

Most women I know who are married into a farming family not only help on the farm, but they also hold a second job off the farm to earn extra cash to make ends meet. They are up at the crack of dawn to finish their farm chores before they marshal the children off to school and then go to do a full day's work for another employer. During the busiest times of the year, such as lambing or calving, farmers' wives often need to take precious 'holiday' from their day job, just so that they can spend all day picking up the slack, doing harvest work or helping out in the calving or lambing sheds.

Marrying into an existing farming family can bring a truckload of heavy expectations, including the fact that you suddenly become an extra pair of (unpaid) hands or that you're immediately responsible for a particular aspect of the farm, like the pet lambs or feeding the youngest calves.

Like many farmers' wives I feel like a single parent: Steve works at least eighty hours a week, and during the summer holidays when the harvest is ready to cut I often don't see him for days at a time. Sometimes I'm near breaking point as I try to juggle my job, my own farm chores and my kids. I'm envious when I see families enjoying a trip to the shops or a day on the beach at the weekend: Steve never stops working, and nowadays he never comes on any holidays with us as he can't leave his responsibilities.

I find working away from the farm distracting and stressful. I know there are many women who love their jobs and are happy dividing their time between the farm and their own careers, but I'm resentful that High House, even though we work *so* hard, still doesn't make enough money to provide us with enough income without relying on a second job with another employer. It amazes me that in the past few generations on both sides of our family, I am the first farmer's wife to work outside the home since 1832. (Note: The exception is Keziah Urwin in 1815 who, after the death of her farmer husband, kept herself and her three babies fed by running a very successful pub in South Shields.)

Farmers' wives not only face the relentless grind of the farm and the endless worries about money but also often have to shoulder the weight of expectation from their partner's families, potentially difficult in-laws and the strains of caring for elderly parents and young children. The stress of splitting every waking hour between farm work, a job off the farm and childcare can be literally overwhelming. A farmer's wife often has to work on the farm, but the business can't afford to pay her a wage, and unless she's a full farm partner she has no say in the eventual direction of the business. Families can be suspicious or resentful of an 'incomer' and can be critical or judgemental of a new wife or girlfriend, chipping away at her confidence and resilience.

When I was diagnosed with generalised anxiety disorder in my twenties I started on antidepressants, and to date I've taken twelve different antidepressants from the newest medications right down to the older, archaic drugs. At first these tablets helped enormously, improving my mood and my social anxiety and insomnia. But as I've grown older they've lost their efficacy, and alongside a slew of horrible side effects, my anxiety hasn't abated but in fact seems to have adapted, warping and distorting into new phobias and obsessions. I know that tablets are immensely helpful for many people, but for me I found it more helpful to cope with my anxiety in different ways.

After both my children were born, I had postnatal depression: an appalling couple of months of grey, flat bleakness where I couldn't eat and struggled with intrusive thoughts of leaving my family, getting into my car and deliberately crashing so that I didn't have to go home. I remember sitting on my sofa with a kind midwife, rocking my newborn baby in my arms, trying to choke down spoonfuls of cereal as she talked gently to me about imbalanced brain chemicals and promised that the black fog would lift. Steve took over all the farm work as I couldn't motivate myself to get out of bed, never mind go outside. He was getting up at five o'clock in the morning, helping to give the baby a feed and then starting a full day of sorting sheep for the mart, or climbing on the tractor to spend twelve hours drilling seed. Being a farmer's wife means that you are often on your own, and in my case I found that very hard, especially with a toddler and a newborn baby. Fortunately, my mum swung into action and visited every single day or took my daughter out to give me time to recoup on sleep. I never, ever want to feel like that again, and my heart goes out to anyone that struggle with depression – my short experience of feeling utterly hopeless was enough to make me understand why people feel that they have no escape. After a change of antidepressants – and the love and help of Steve, Mum and Dad – the fog lifted and I began to enjoy being a new mum. In a few more weeks I felt strong enough to strap the baby to my front in a sling and between feeds I would toddle out onto the farm to see the ponies and the sheep. It felt amazing to be back outside and reacquaint myself with all my favourite spots on the farm, and to pat the horses and see how the lambs had grown. I remember wobbling down to the ash tree that overlooks Sparrow's Letch, the little stream at the bottom of the front field. Still feeling sore from the C section, I carefully lowered myself down to sit with my back against the roots of the old tree, closed my eyes and tilted my face to the sky, letting the rays of warm sun play across my eyes and nose. My baby lay kicking on a blanket on the short grass, and at that moment it felt like I'd finally woken up: as if the world had begun to turn again and I had slotted back into the right time and place. When you have been so low the first glimmerings of hope and happiness feel like your prayers have been answered, and I know that the farm had a part in making me feel like myself again. Life fell back into a routine and, with a toddler in tow and a baby in a sling I began to help Steve. Mucking out pens and feeding sheep with very young children is an exercise in patience and ingenuity, especially when surrounded by dangerous machines and unpredictable animals. But with the help of a pram, a sling, tractor seatbelts and the occasional help from a small pony, we managed with everyone emerging (mostly) unscathed.

There is something inherently anxious about being a farmer, and personally I think it boils down to the terrifying lack of control that farmers have over their livelihoods. Bad weather, storms, heavy rain or droughts can mean the difference between making a profit or breaking even. Government rules and regulations, massive increases in seed or fertiliser prices, the threat of animal disease and drops in milk, beef, pork or lamb prices can have a devastating effect on your business. And the survival of the farming family is tied to the success of the farm; it is a tremendous weight for anyone to carry. If you're a big landowner and have a comfortable safety net in the bank and extensive land and property holdings, plus estate managers to oversee your various businesses, then I would imagine that farming in the UK must be a great way to earn a living. Unfortunately for the rest of us, trying to make a living whilst bringing up a family can be extremely worrying, especially if outside events begin to impact on the health of your family.

During the first wave of Covid, my anxiety began to spike, and my intrusive thoughts started to spiral. I remember very clearly reading an article in February 2020 about escalating

Covid deaths in the small town of Bergamo in Italy and looking in horror at the accompanying photos of ranks of elderly men lying prone in the Italian ICU, unconscious and intubated as the virus ravaged their lungs and internal organs. At first, my parents didn't realise the severity of the risk, and I struggled to make them understand that cancelling their upcoming cruise down the River Rhine was absolutely necessary. Eventually the rising Covid death rate persuaded us all, and we retreated almost thankfully into lockdown. I was so anxious that I couldn't sleep or read, and although it sounds excessive in retrospect, I began to make plans about how I was going to feed my children and keep them safe if society started to break down. The only thing I found even slightly helpful was the fact that workload over lambing was so heavy that I didn't have much time to think, and pure tiredness helped banish any very upsetting intrusive or ruminating thoughts. During the second lockdown, Covid restrictions became less worrying, and I settled into a routine of online shopping, mask wearing and handwashing, and I did feel a lot safer and less apprehensive after I had my vaccinations. It's amazing how adaptable the human brain is and what was once terrifying can become almost mundane and an everyday routine.

Why am I telling you all this? I promise that it's not for attention or an attempt to curry sympathy. I've always been open about my struggles with anxiety, and after writing about my mental health issues I received an armful of private messages from individuals who were keen to tell me their own battles with depression and anxiety. There is something comforting in knowing that other people are going through the same things as yourself and under resourced with a lack of qualified practitioners and months-long waiting lists.

I know people bang on about the benefits of being outside, but a short walk down to our wood – my feet sinking into the deep, green turf as I amble past our peaceful sheep flock – calms my jumping thoughts and negative self-talk. Reading books also helps. I read all the time: everything from fantasy and fiction to history and more. Reading is like watching a movie in my mind, and I can visualise the characters and the settings as vividly as watching them on a screen – it's pure escapism. And finally, seeing my friends pulls me out of dark places. I don't have many, but the ones I have I trust implicitly, and they can cajole me out of my bad, self-obsessed mood and make me laugh so hard that I snort tea down my nose.

When the depression or anxiety becomes too much, I turn to my GP, who has limited options but will try their best, recommending a new tablet, an increase in dosage or another go at therapy. I dream about proper mental health hubs, organised like a hospital, where you can just turn up and you'll be triaged sensitively and quickly, and receive timely, effective and professional help either in-house or as an outpatient. So many problems with individuals in society are down to their unravelling, untreated mental health and I wonder if in the future any similar solutions will be put in place or if the situation will need to reach crisis point before enough government funding is released to provide effective help.

Listening to previous experiences of my mum's generation and how they overcame their own struggles with anxiety and postnatal depression gives truth to that popular phrase 'this too will pass'. My mum explains how she managed difficulties and anxieties about her family and the future. She encourages me to look at my life in perspective, to avoid catastrophising over simple problems and to see that things are not as bad as they seem. The rhythm of the farming year also helps: no matter what happens, the season turns, and summer comes around once more bringing sunshine and green fields. And no matter the time of year or weather, High House always has something to offer – I can find just as much beauty in the bare branches, icy streams and bleached grass of winter as in the more obvious splendour of oak trees in full leaf or wheat fields rippling golden in a summer breeze. Sometimes all you can do is just hang on, wait for the rain to stop and look forward to the sunshine.

CHAPTER EIGHT

Horses and Dogs

The very first modern sheepdog was born in Northumberland in 1892, about twenty miles, as the crow flies, to the west of our farm. Shepherd Adam Telfer bred a pup called Hemp to his black sheepdog bitch Meg and his tri-coloured herding dog Roy. Hemp apparently had intelligence, sheep-herding instincts and working abilities that amazed the local district, and he became a much sought-after stud dog, so much so that it is said he sired over 200 descendants. Over the century, Hemp's offspring spread across Northumberland, and his progeny eventually became celebrated as the pinnacle of sheepdog breeding; our collie Mavis is probably descended from the very same line. At High House Farm we commemorated this very first collie with the original beer produced in our brewery: an amber-coloured best bitter named Auld Hemp. The next beer we brewed was named Nel's Best after our previous collie, Nel, who was a beautiful tri-coloured bitch with a fondness for cheese scones and ramblers' packed lunches.

Steve has always owned a working collie dog on the farm to help manage our sheep flock. Mavis, our current farm collie, is a whirlwind of energy and fur with a sleek, black coat, a shiny nose, deep-brown eyes and a wonderful, feathery tail that curls up over her back. She has a one-track mind, and her entire life has been focused solely on herding and chasing sheep. From the moment she wakes up to the moment she flops down to sleep she is obsessed with watching our sheep flock, lying for hours by the front gate, nose on her paws, beadily watching the grazing animals. She's also very bonded to Steve, and whilst she'll put up with my company if there's no one else around, as soon as she hears the quad bike or the telehandler engine she's off like a shot, skidding over the grass to see what Steve is doing.

Mavis is trained to gather the sheep, keeping herself opposite to the dog handler, moving them in any direction that the handler needs them to go. She's incredibly bright and often seems to know what I want before I ask her. She knows lots of commands such as 'come by' (to run left) and 'away' (to run right), 'fetch on' (to bring the sheep up to the handler), 'look back' (to make sure she spots any stragglers), 'here to me' (to move towards the handler), 'that'll do' (to tell her the work is over) and of course the ubiquitous 'lie DOON!', which is usually bellowed at top volume when Mavis becomes a bit overenthusiastic about the whole business. Unfortunately this enthusiasm sometimes gets the better of her patience, and if she's bored and our backs are turned she has been known to go and collect a batch of sheep herself, then stand proudly at the back of the flock, panting happily as the sheep cluster in a bewildered huddle by the front gate.

Mavis also is very much an outdoor working dog and is disdainful of the life of a pampered indoor pooch. At first, we encouraged her to sleep in our house, giving her a warm bed in the kitchen, but she spent all of her time slumped on the doormat whining and wanting to go outside. Now she sleeps in a warm stable surrounded by lots of cushions and straw, and she seems much more settled.

She isn't trained to walk on a lead and doesn't understand the point of walking at heel. Instead, she trots along in front, nose to the ground, watching beadily for any sheep in the distance. Mavis does enjoy a dog walk but is very nervous off the farm, so I tend to just take her down through our front field, persuading her to leave the sheep behind and investigate our ten-acre wood. Then she becomes like every other pet dog: tail in the air, nose quivering as she bounds through the undergrowth, flushing out rabbits and rolling in revolting smells.

She's brilliant with people and children but very overprotective when it comes to other dogs, so I've often had to grab her collar when we're herding sheep to stop her spotting some innocent doggy bystander and trying to obliterate them for daring to look in our direction. We're still recovering from the time she glimpsed a mini dachshund in our farmyard, bumbling along behind its human owner, decked out in a matching collar and lead and wearing a tiny, blue bow tie. We were moving sheep, and Mavis and I both clocked the tiny pooch at the same time; I literally had to lunge for her collar as her hackles came up and she tried to bound over to the trespassing dog, teeth bared in a terrifying snarl. The mini dog cowered to the ground, the owner nearly had a heart attack, and I spent a long time apologising, trying to pat the tiny dog and surreptitiously adjust the now very muddy bow tie.

Mavis is in her prime of life, seven years old, with a glossy coat and bouncy gait and is still just as obsessed with sheep as when we bought her. We're hoping she'll be part of our family for a very long time, and as her pedigree is impeccable we're considering finding her a similarly aristocratic boyfriend so that she can raise her own litter of puppies.

As well as dogs, the farmers at High House have always kept horses and ponies. On the old map dated 1832 there is a seven-acre field clearly named as the 'Horse Whins' where they pastured the horses who pulled the ploughs and carted the straw, hay and muck ('whins' is another name for gorse bushes, which still cover the bottom half of the pony paddock). In the farm buildings, you can still spot the original stables with bluestone-cobbled floors, remains of tethering rings, bolts from old, loose boxes and horse stalls, and cobwebbed hatches in the roof leading to a very dusty hay loft, which must have been rammed full of feed during the colder months. I've found other hints of past equestrian inhabitants including heaps of old horseshoes, from tiny sizes that would have fitted donkeys and Shetlands up to huge rusty slabs of iron that must have been hammered onto plough horses' hooves. The sheds have piles of old tarnished horse bits and leather horse harnesses, corroded chains and discoloured plough parts that were all dumped in a forgotten corner when they broke or wore out. One huge horseshoe was lost at the roadside in the 'Horse Whins' field, and over the years the spiky blackthorn hedge has actually grown into and around it, so a branch is wrapped around and around the curved metal shape, absorbing it into the bushy stem. From the size of the shoe, the horse must have been over eighteen hands high – a huge beast with the strength and muscle to pull a plough or harrow through the heavy soil of our farm.

My mum and dad tell me that even as a tiny toddler I loved horses. I would press my nose against the lounge window to stare at the ponies and riders who clip-clopped down our street to reach the beach at the bottom of the road or insist on spending time with every horse that we met on our walks, taking ages to pat their necks and stroke the soft noses, testing the patience of my brother who didn't share my fascination with anything equine. I loved the possibilities of horses: the excitement and freedom when cantering a pony across a field or the exhilaration of jumping a hedge. I wanted to become a horse's best friend: to earn their trust so that they rested their soft nose on your shoulder or snuffled into your hand. Every car or train journey was spent imagining myself on horseback, jumping the fences and hedges that raced past the window, and I read every horsey book available, drinking down 'Silver Brumby' or 'Phantom Horse' stories whilst waiting for the next instalment of *Black Beauty* to appear on our TV. At seven years old I started lessons at Murton Riding School, and whenever my parents could afford the ten-pound fee I would turn up early to the stables so that I could pat each pony in turn and inhale the horsey smells. Riding was the highlight of my week, and I was absolutely obsessed with all things pony; but living in a terraced Victorian house by the seaside with completely animal-averse parents meant that I never managed to own my own horse. I was a good little rider though, neat and precise, trotting around the school and doing serpentines and diagonals. I lost my nerve a little when I entered my teens and I rode a pony that had a reputation for running off with a rider. The first half of the hack went well, but when we turned for home the animal suddenly took off, bolted down the pavement, charged straight across the road, missing cars by inches, and then jumped a hedge, throwing me off in the process. My riding became a little tenser, and I started to lean forward with nerves rather than sitting tall and straight.

As I grew older, I grabbed whatever riding lessons or hacks I could find: visiting stables in Leeds and London, taking occasional one-to-one lessons on the lunge or over small jumps. I never rode enough to gain any sense of progress and it was only when I met Steve that I was able to indulge my childhood dream and own my first pony.

My very first horse was a 14.2 hands Welsh Arab cross called Henry who to this day still retains the dubious honour of being the most neurotic animal I have ever met. He was a deep chestnut colour, all glossy like a conker with a nut-brown mane and tail, but his handsome looks belied the fact that he was constantly one hoof away from a nervous breakdown. Henry was scared of everything – including but not restricted to: lines painted on the road, pigs, sheep, horses he didn't know, plastic bags, buckets, bicycles, other people, trees and crisp packets. Every hack would start off with me slinging a leg over his quivering back as he swivelled his eyes in terror before shooting off at a smart trot, shying at hedges and gates as he snorted up the road. Every ride out would culminate in him spotting a lurking monster (aka a wheelie bin) in someone's drive, bucking in terror until I fell off and he triumphantly galloped back to stand outside his stable. Steve used to have a bet on how long it would take before he saw Henry careering down the road, with an empty saddle, his stirrups swinging as he skidded back into our farm drive. I used to enjoy galloping up a hill beside my parents' house as it was steep enough to ensure that Henry was too knackered to do anything silly; but eventually, after falling off him three times in one morning, I decided the phobic pony had to go. He was eventually sold to a very good rider who turned him into a fast and brave three-day eventer.

Then there was Bob, the hugely rotund black cob with fluffy black legs, who the owner promised was 'quiet as a mouse'. Bob's piece de resistance was planting his four feet and refusing to leave our farm drive, and walking backwards every time I tried to get him past our gate. When I *did* manage to get him out on a hack, he trotted faster and faster until I was as tense as a bow string, bridging his reins on his neck in an attempt to stop him crossing his jaw and bolting back to his stable. One winter's day, the local hunt forgot to tell us that they were meeting near our farm, so instead of shutting Bob safely in his stable I'd left him out, peacefully grazing in his field. I think it was one of the best moments of his life. According to a hunt follower friend, Bob spotted a large group of horses and riders trotting past his fence and made the snap decision to join in, even though he hadn't actually been invited. He lumbered into a gallop – speeding up as he approached the fence and leaving huge hoof divots in the turf – and launched his enormous bulk up into the sky and over the gate. An amused hunt member brought Bob back home, leading the sweaty pony on a spare lead rope as he high-stepped up the road in excitement, his eyes gleaming and tail lifted in pride.

Finally there was Nemo, a piebald 13.2 hands cob who was slightly too small and had an amazing trick of cantering a corner, lifting his back and dropping his shoulder so that I slid off sideways and landed in an undignified heap on the ground. His entire aim in life was to eat as much food as he possibly could, and every hack culminated in him taking the bit between his teeth and ending up nose down by the side of the road, ignoring my flappy kicks and screeches as he determinedly ate through three tons of roadside greenery.

In reality, all these ponies were far too 'sharp' for me, and I didn't have the confidence or skills to ride them properly. After each failed horse-owning experiment my confidence withered until I vowed I would no longer try to ride until the kids were grown and I had more time and support. I still long to own a horse, but if I try to get back into the saddle my anxiety kicks in, I shake and wobble and, embarrassingly, sometimes burst into tears. Once the children are bigger I'm determined to join the world of horse ownership once again, although I tease my kids by saying I'll buy a driving donkey and tub cart so that I can go to pick up the newspaper from our local shop. My family is horrified by the mere thought of me trotting around the local lanes behind a moth-eaten mule and has so far managed to veto any riding lessons or future pony shopping.

Candy, our little 11.2 hands Shetland cross pony, was originally bought so that my children could learn to ride, and she must be the most famous horse that has ever lived at High House: she's a social media star with her own hashtag on Twitter. Lots of people seem to enjoy her sassy personality and general lack of cooperation. She regularly receives packets of mints and Polos in the post.

Candy is now a very elderly thirty-two years old and has definitely slowed down over the last few years. Her front legs are stiff with arthritis, and her stride has shortened so that instead of launching into her usual canter or gallop she only manages a short burst of shaky trot, and even then only if absolutely necessary, such as when spotting her feed bucket or running away from the vet. She has COPD and Cushing's disease, and she takes one sachet of painkillers a day; she needs to be sedated for the blacksmith and has dust-free hay and rubber matting in her stable to stop her wheezing and to protect her sore joints. Even though she's not as spry as she once was, Candy still has a gigantic personality and makes it very clear when she doesn't approve of any request, no matter how much bribery is involved. She hides from the vet and isn't entirely convinced about Steve as she knows they both like to prod her with needles or squirt foul-tasting worming paste down her throat. Candy's greatest goal in life is to eat as much as possible in the shortest timeframe, and her physique demonstrates how successful she's been so far. Conversations with my long-suffering vet go as follows.

'I think she's lost a bit of weight,' I say hopefully every autumn, as Candy sulks at the back of the stable, eyeing Dave the Vet with suspicion.

Dave runs his hands over Candy's neck and back and prods her thick, white coat.

'She's still quite a chonk, but she's carrying it mostly behind her girth.'

We both stare at Candy's stomach, which has slipped so that it curves out like a displaced water balloon behind her waist.

She basically looks like a small ball balanced on top of four skinny legs.

'She must have had lots of foals,' says Dave charitably, 'or maybe it's just her natural shape. Anyway, you'll still need to keep her in a grazing muzzle and on a bit of land that doesn't have much grass. Too much grazing and she'll get the dreaded laminitis. Is she eating that Top Chop Zero feed I recommended?' (Note: Laminitis is a nasty syndrome that usually affects small, fat native ponies who eat too much grass. The sugars in the grass make their feet sore and inflamed. In bad cases the foot bone can rotate and even drop out of the bottom of the hoof.)

I shrugged my shoulders non-committedly. 'It didn't really work that well.'

On the back of the Top Chop Zero bag there is a long-winded description of the contents, explaining that the sack is full of an 'ad lib no sugar roughage feed that can be fed to overweight equines' and 'has been created specifically so that your horse will pick slowly at it during the day'.

I dragged my boot backwards and forwards in the straw at my feet. 'I filled a bucket full of Top Chop, gave it to Candy, left to drink a cup of tea and when I came back, she'd polished off the entire bucket and was looking for something else to eat.'

Dave rolled his eyes whilst Candy gave a snort and carefully investigated her empty feed bowl to see if there was any remaining dinner.

'And she scrapes her grazing muzzle off every time I stick it on,' I continued, staring balefully at the small pony who was now investigating Dave's pockets in the hope of an extra-strong mint.

The grazing muzzle is a clever headcollar which fits over a pony's face and is attached to a sort of perforated bowl that goes over its muzzle. It's supposed to work by restricting a horse's grass intake, allowing them to graze and mooch about a larger field, but not eat tons of grass. In practice, I often find Candy with her grazing muzzle hanging off one ear, or she's missing it entirely as she's spent all night carefully levering off the offending item against a sticky-out section of the dry stone wall. I've lost two of them in her paddock. She's probably buried them somewhere.

Dave patted Candy on her neck. 'She's one of the most bloody-minded and clever ponies I've ever met. Just do what you can to keep her from turning into a butter ball. I'll come and see her again next year.'

Candy is best known for her acting skills, and you've never seen such dramatics until you've seen the thirty-two-year-old pony turned out into a field full of snow. She's very good at finding the best spot to stand and look lonely, dismal and half starved, whilst pretending her very expensive, thick field rug is letting in the rain. I turn her out on winter mornings shouting, 'You'll enjoy it! It's good for you! Go and find something interesting to do!' as if she is going on some terrible horse school trip, and then she stands by the gate, in her duvet-like rug, with the same morose expression as Eeyore. No matter how often I explain to her that her ancestors lived on windswept mountains and bleak moors and enjoyed a diet of moss and twigs, she ignores me and sucks dramatically on a thistle pretending to be hungry, whilst waiting for her next bucket of very expensive veteran horse feed.

Candy is also famous for being dead. She's a very convincing actress, and I've lost count of how many times passers-by have stopped me on the farm drive to tell me sadly, 'I think your pony has died.' She chooses a spot easily seen from our footpath and flops down to sleep on her side, legs out straight and her huge belly sticking up as if she's gone all bloaty and is rigid with rigor mortis. Last year she went one step further and adopted her own personal crow, a kind of living prop, that perched on her rounded side and enjoyed hunting for bugs in her thick coat. From a distance, when Candy rolled on her side and fell asleep with her crow sitting on her belly, she looked like she'd been dead for a number of days and the bird was helping himself to some pony-sized dinner.

She's not a 'cute' pony or one that you could leave unattended with a group of kids; she's too wily and cunning with a strong streak of self-interest, but I love her all the more for this. Farm tours are embarrassing as I usually promise each group a chance to see our 'family-friendly farm pony' as if she was a major attraction. When we arrive in the field, the visitors blanch at the sight of a small pony flat out in the sunshine – stomach distended, eyes rolled back in her head – who ignores any of my attempts to try to haul her to her feet. It's very awkward. I often have to persuade people to pat her whilst she's still lying down and letting out the occasional thundering fart. The little pony seems to want everyone to know that she's being *forced* to join in my tours and be patted by small, squeaky children whilst listening to my repetitive jokes.

Candy is very wary of children, and I have to keep a tight hold of her headcollar when a small person comes over to give her a pat. Left to her own devices, she's likely to bite or lower her head and thump them hard in the chest. She hates being saddled up, and tightening any girth around her large tummy leads to her biting in panic. She's also a little head shy and refuses to accept a pat between her eyes or a stroke on her nose. But I can excuse these signs of temper as we're not sure what behaviour she faced at her riding school, and it is the nature of small children to yank up girths or tug on ponies' mouths when learning to ride. She's now fully retired and is living her best life, mooching lazily around her paddock or begging for treats over her field gate. She still enjoys a tummy scratch and stands for ages, lifting her right leg in supplication to persuade me to continue scritching her itchy coat, turning her head to the sky and grunting in contentment as she tries to reciprocate by grooming my hair with her teeth.

She shares her paddock with some elderly sheep and has seen the quad bike many times – she's often been led behind one – but if the wind is in the East and she's in a particular kind of mood she likes to pretend she's frightened and turn it all into a game. She's very good at escaping, and whilst she doesn't have an amazing turn of speed, she makes up for it in pure devilry.

One morning Steve opened her gate and rode round the paddock on his quad, sending Mavis out in front to round up the small sheep flock so that he could trim their feet. Usually, Candy just completely ignores the quad and the dog, but on this particular day she was feeling sprightly and joined in the small group of elderly ewes, trotting smartly along up to the gate. Steve tried his best to turn her back, and Candy suddenly decided that for some unknown pony reason today was the day she was going to charge *through* the small flock, splitting them three ways and chasing them all out onto the main road. Mane and tail flowing behind her and ignoring Steve's exasperated shouts, she wobbled down our main road, herding the sheep in front, bucking and farting with excitement, sidestepping cars and leaving long skid marks on the tarmac. After a while the sheep grew fed up and dived into the hedgerow, and Candy slowed to a walk and began to munch on the grassy verge, making sure that she pointed her large bottom into the passing traffic to ensure maximum driver consternation.

I stormed down the highway, waving her headcollar and ignoring drivers who wound down their windows to tell me about the 'white pony' they had seen prancing past their cars.

Candy saw me coming, correctly read my facial expression and whipped round, trotting smartly down the lane towards Newcastle. Fortunately, her sore knees and general lack of speediness meant that I managed to catch up and, throwing the lead rope round her neck, brought her to a screeching stop, leaving a long, black slide in the tarmac. She thrust her head into the thick roadside grass, pretending that it was all a mistake and she wasn't in trouble at all.

The drivers cheered as I dragged the small recalcitrant equine back up to the farm, bouncing fatly along on the end of a rope and chewing thoughtfully on an enormous sprig of cow parsley.

'Oh, she's GORGEOUS!' yelled a small child in the back of a car.

Candy was massively proud of her escape attempt and showed off in front of the crowd, snorting and curvetting her neck like a dressage horse, stamping her tiny feet and making faces at the drivers peering out of their windscreens.

Steve was apoplectic with rage.

'That bloody pony!' he bellowed as she appeared around the corner. Candy ignored him, studiously avoiding his gaze as she stared across the farm and finished off her roadside snack. 'I've got half of the sheep down in next door's place and the other half on their way to Hexham!'

It took us the best part of the afternoon to gather up the strays. A protesting Candy was shut into her stable and given some painkillers for her joints and a net of hay whilst we sorted out the rest of the sheep. She spent all afternoon reliving her glory, snorting and whinnying with suppressed excitement until she fell asleep with her head in her feed bucket.

Even though she's retired, one day a year the fat pony is pressed into action as a key part of our family brambling day out. The hedgerows at High House Farm usually brim with blackberries, with thick bunches of black, glossy berries sprinkled across the spiky branches. We pick a day, and the kids and I and my mum and dad (Steve is almost always in the middle of baling hay) meet at a preorganised time, with plastic containers and gloves, ready to pick the autumn crop. I had the brilliant idea of using Candy as a beast of burden to carry the heavy baskets of fruit, so one afternoon I created a Heath Robinson type of saddle pannier using two plastic buckets, her saddle and plenty of baler twine.

The pony seemed to enjoy her day out, ambling along with two buckets precariously hanging on either side of her saddle, tootling along the road to the chosen stubble field and grabbing large mouthfuls of verge-side grass.

Once in the field, we closed the gate and took off her lead rope, and she mooched along behind us, sorting through the hedgerows for the most luscious mouthfuls of grass and the occasional blackberry leaf. She would stick close to us but sometimes wandered off across the field to peer over the fence next door to stare at the cows grazing on the neighbouring farm.

This year, the dry summer meant that the fruit crop was particularly good, so we filled her Tupperware boxes up to the brim with enormous blackberries and she stood patiently, accepting the occasional mint sweet until we were ready to go home. By this time we'd been brambling for a good few hours, and my mum had accidentally fallen into the stream at the bottom of the field trying to reach a branch laden with blackberries. Dusty and soggy, we slogged our way back up to the road, and I glanced behind to check Candy was following behind.

The small pony had stopped a few feet away and had twisted into a pretzel shape so that she could fit her muzzle into one of the buckets hanging off her saddle. I saw her greedily gulping down huge mouthfuls of fruit.

'No!' I shouted, making a grab for her headcollar.

Candy jerked up in surprise, took one look at my face and promptly went into reverse, whipping around so quickly that one of her buckets flipped upside down, dumping all the blackberries onto the stubbly grass. She trotted off to the far side of the field, carefully keeping one eye behind her, and started eating sneaky mouthfuls from the remaining bucket still tied to her saddle.

'Oh god,' I said, bending down to start picking up the spilt blackberries and stuffing them into a spare plastic bag. Candy was now at the far end of the field, head down in the grass, deliberately ignoring our entreaties to trot back towards us.

'Oh just leave her,' sighed my soggy mother, 'she'll come over when she sees us leaving the field. She won't want to be left on her own.'

Clutching our squashed fruit and covered in purple juice, we trekked up the field back to the gate. Candy raised her head and watched us suspiciously as we crested the hill, waiting for the very last moment until she finally whinnied noisily and cantered creakily towards us, skidding to a halt in front of the open gate. I clipped on her lead rope and she nuzzled me affectionately, her muzzle stained with purple juice.

Half of the fruit had gone, but Candy was very proud of being part of the blackberry-picking team, curving her neck and trotting back home at the end of her lead rope, huffing with excitement and pleasure. We untacked her and turned her out into the field, where she promptly rolled, adding big, dusty patches to her purple-stained coat. We managed to save about half a pound of brambles from the day, enough for a large apple-and-blackberry pie. Candy is now banned from any future bramble-picking days, and she stares mournfully at us as we walk past her gate, laden down with buckets and Tupperware brimming with fruit. I still give her a good few handfuls though as I go past.

Although Candy has my heart and I love her to bits, there have been other ponies that have won my affection.

A few weeks before Steve and I were married, I'd pulled into a lay-by on the A1 to check my phone on my way to work. Glancing up through the windscreen, I saw a couple of ears poking above the weeds at the side of the road. On investigation, they belonged to a tiny, dirty brown Shetland pony who was so small she was almost hidden behind a bank of nettles and long grass. She was emaciated and lice-ridden and stood with her head down near her knees, her nose and eyes crusted with flies.

As I approached, she pinned back her ears, bolted to the far end of her tether chain and turned her backside towards me, threatening to kick. There was no water and the grass had been cropped down around her to a bare, dusty circle.

I rang Steve.

'There's this pony I've found by the road…' I tentatively began.

'Oh god,' he said, as if I *regularly* trawled the countryside and smuggled ponies onto the farm.

'What shall I do? She's filthy and thirsty and terrified of everything.'

Eventually Steve persuaded me to ring the local branch of a rescue society, who promised to come out and investigate.

I sorted out a tub of water and asked the staff from a nearby petrol station to keep an eye on her, and I reluctantly left the poor little thing on its own again. That evening, I got a call from the rescue society who told me that the pony had been dehydrated, that there was evidence of her being beaten, and asked if, once she was recovered, I would like to come to see her.

Of course I would.

A few weeks afterwards Steve collected Gladys the pony from the rescue centre in his cattle trailer. The small pony took a bit of persuading to walk up the ramp, but eventually she stepped in

reached the grand old age of thirty-five, and after a drenching rainstorm caught a cold that turned into a bad case of pneumonia. I stabled her in the sheep pens and the vet gave us a big bottle of 'drench' – liquid medicine that had to be poured into her mouth every day. Steve took on this unenviable job and braved kicks and bites from a very cross Gladys as he tipped up her head and poured in the purple medicine. When she recovered a little and the weather was warmer, we turned her out in the small paddock and she pottered around, wheezy but content. One morning I went out with her usual feed of veteran pony nuts and vitamins and found her dead: flat on her side, right in the gateway entrance, her legs out straight and her body cold and stiff. According to the vet her heart had just 'given out', and she had dropped down, dead before she'd even hit the ground. In the short time we'd had her, I hoped that she'd come to understand a little how humans could be trusted and experienced enough love and care to make up for the cruelty and neglect she'd faced in her younger days.

In more recent times, Candy has enjoyed a variety of pony friends, all loaned to us from nearby farms for the few months in the year when they're not needed as riding ponies. These little Shetlands and Welsh ponies keep her company and are her partners in crime. We've had Snowy, Tinkerbell, Noisy and the current incumbent Jingle the Welsh Section A pony, who is on loan until the spring. Jingle is a tiny bay mare with fluffy ears and legs, whose piece de resistance is jumping into next door's field to hoover up leftover sheep feed. Candy isn't athletic enough to jump anything but watches in excitement as Jingle bounces over the gate and gallops through the flock of sheep, scattering them to the four winds so she can settle down to eat their dinner. It seems to keep the ponies occupied and watching Steve charging after the small pony to try to drag her back into her own field is always very entertaining.

CHAPTER NINE

Brexit and Piggies

My grandfather used to say that once in your life you need a doctor, a lawyer, a policeman, and a preacher. But every day, three times a day, you need a farmer.

Brenda Schoepp, brendaschoepp.com

After the brewery and wedding business passed onto another couple, the income we received from renting the farm buildings to the new owners just covered the monthly repayments on the mortgage for the original diversification, leaving little or nothing for day-to-day expenses. The farm made only a marginal profit, and I was working part time as well as looking after children. Steve made the decision to try to work three days a week at the local agricultural merchants to bring a little extra money to our family. He got up every morning at four thirty to check the sheep and sort out any problems, then came back to get ready and go to work. He worked an eight-hour shift in the merchant's office, then once he came home, grabbed a quick dinner and was straight back onto the farm, hopping onto the tractor to plough fields, sow crops and look after the animals. After a while he grew more and more stressed and began to catch every cold or bug going, often only managing three or four hours of sleep before he'd push himself into the shower and go back to work. His employers were lovely folk, but even they became worried as his weight dropped and he began to look drawn and grey. When lambing season came around, Steve looked almost skeletal and every time he sat down he almost immediately fell asleep. One morning, he got up early to check the ewes and lambs and fainted clean away in the sheep shed. I only found out when he came back early that evening and showed me a bruise on his forehead from where he'd bashed it on the side of the feeding trough.

'This is no good,' I said to him as he crashed around the kitchen, trying to grab himself a snack before he went out onto the tractor again. 'You're going to make yourself really ill. You never take a break. When was the last time you did anything other than work?!'

'What do you want me to do?' he snapped in return. During the past few months he'd been almost continuously grumpy with an easily triggered temper, and his irritation was palpable. 'We need the money. You need the money for the kids. You're always going on about how skint we are, so I can't stop working.' He grabbed a packet of crisps and a Diet Coke and started to pull on his jacket and welly boots. 'I need to roll the bottom field. I'll be back later on.'

'You're going to kill yourself if you carry on like this!' I shouted at his disappearing back, as he stomped out of the house banging the front door behind him.

When we'd both cooled down we had a proper chat, and Steve admitted how stressed he felt trying to juggle a full-time farming job plus a part-time office role. I forced him to go to his GP and explain what was going on, and with support Steve decided to hand in his notice at the agricultural merchants. It was just impossible for him to carry on working two jobs, never mind remembering to eat properly, and he was hardly sleeping.

We decided to have a meeting with our bank manager to see how else we might be able to diversify and bring more income into the farm.

One cold morning in December I stood in our kitchen, clutching Cinders the cat around her fat middle, nervously stroking her head whilst listening to the bank manager who sat at our kitchen table.

The bank manager raised his eyebrows as he looked over the previous year's accounts.

'You have to use what you've got. Make your assets sweat some money,' he said.

I coughed nervously. *Sweat some money?*

'Why don't you look at free-range chickens or high-welfare pigs? They both bring in a fairly decent income.'

'How much would we need to borrow to set them up?' I asked. Cinders was becoming grouchy with all the stroking.

The bank manager sighed and leant back, placing his biro carefully on the kitchen table.

'You're looking at around £600,000 for the pigs and over a million for the chickens.'

Steve choked on his mouthful of tea, whilst I gave Cinders an involuntary squeeze and she screeched to be let down.

'It's a big risk,' he continued, 'but you need to do *something* as the sheep and crops just aren't bringing in enough, and your overdraft is *right* on the edge.'

We nodded like puppets as he raised his eyes over the profit and loss statement. 'I mean, you could rent out your fields to other farmers and go out to work, but…' He trailed off and looked up at Steve.

'No,' said Steve immediately, 'I don't think I'd cope with someone else working the land.'

The bank manager nodded and clicked his ballpoint pen. 'Right. Well, you go and do a bit of research and create some figures, and I'll take a look and see what I think.'

As part of our fact-finding mission we went to visit a large egg producer down in County Durham. Their setup was immensely professional, and Ian, the farmer, gave us a rundown of the business, explaining about his 400,000 free-range chickens, how the system worked and everything they had to do before the eggs could be sold to supermarkets and consumers.

These chickens were properly free range, in ten football-pitch-sized sheds with perching bars, massive sawdust floors and sides that opened up to allow the hens to wander outside, each shed sitting in its own forty-acre fenced, grassy field. We had to boot and suit up, carefully disinfecting our boots before we were allowed in to see the birds, who were packed into the shed, their clucking at deafening levels.

'Why are they all inside?' I bellowed to Ian, over the noise of the thousands of chickens clucking and squabbling in the massive barn behind.

There were chickens as far as you could see, crammed onto the perching bars, scuffling among the sawdust, pecking at feed and water or snoozing as they roosted against the side of the shed. The sides of the barn were open, and it was a glorious day, but I could only spot a handful of nervous-looking chickens pecking at the ground outside.

'Bloody red kite,' said Ian, pointing up into the clear sky. 'They keep picking off the birds so they're all scared to go outside.' He sighed. 'I'll have to fence over the top to stop them, but it'll cost a fortune.'

I realised that the local wildlife was using the laying hens as their own personal larder.

'I found a fox in with them once,' said Ian with a shrug. 'It'd crept in under the fence, through the open sides, and was fast asleep under the egg conveyor. I think it was a young 'un and was just overwhelmed by the amount of dinner it found.' He grinned, showing his teeth. 'It didn't half shift though when I appeared. Bloody thing was like a streak of red lightning. We shored up the fences and nowt has got in since.'

The hens clucked behind him, eggs dropping down into the sawdust or onto the soft conveyor belt. Suddenly machinery clanked and groaned, and the belt started to move, the eggs riding along it, wobbling from side to side. Ian took us into the next section of his farm, where the eggs, of all different shapes and sizes, appeared through a curtain of rubber strips before a team of workers – all in hairnets, boiler suits and masks – carefully picked up the eggs and plopped them into huge vats of water.

'They're cleaning them, see?' said Ian. 'Washing off any of the muck or feathers. Supermarkets don't like it if there's any speck of dirt on them as they won't sell.'

He pointed to a row of half-filled egg boxes behind him. 'All the ones that aren't perfect are put in there, and we sell them at a discount. Even the ones that are a bit smaller, they can't go to the supermarket, so we sell them direct to catering companies.' I stared at the rows upon rows of eggs, white and brown and all the shades in between. They all looked the same size to me, although after looking carefully I could see that some were slightly smaller or had a hairline crack in the shell.

I was impressed by the scale of the operation, the number of quality checks on both chickens and eggs and how passionate Ian was about the business. He took us to his tiny office and showed us a sheet of figures.

'This is how much the shed cost in total, with all the equipment and computer bits,' he said, pointing at a frankly mindboggling number at the bottom right of the page.

'And when did the business start turning a profit?' asked Steve, stroking his chin as he stared at the piece of paper.

Ian leaned back in his office chair and stretched his arms above his head. 'After around three to four years,' he said, 'as long as you can organise the finance in the first place.'

On the way back home in the car, Steve and I discussed how much money we'd have to borrow if we wanted to set up a similar operation. The costs were astounding, and the long gap between paying for the set-up and turning a profit worried me. 'I can't see the bank lending us the money,' said Steve, his hands on the wheel. 'I don't think we have enough assets to offset the mortgage. Maybe if the farm was bigger, or if you had a full-time job.' He turned his head to look at me as I chewed my lip. 'It's a hell of a cost.'

He was right.

The bank took one look at our proposal and turned it down flat. Our bank manager was apologetic on the phone. 'Sorry guys, the figures just didn't stack up.' He blew out a breath. 'I tried to push the loan through, but the finance committee wouldn't listen.' He paused for a moment. 'Why don't you try looking at pigs instead? The initial cost isn't as high, and it might be a better proposition for you guys.'

Meanwhile Steve and I weren't just looking at the possibility of setting up another farming business. We'd also decided to look at other farms for sale, with the idea of selling up High House, clearing our substantial debts and buying another farm in Scotland where the cost of land was slightly cheaper. It wasn't a decision we took lightly, as Steve loved his farm, had lived his entire life at High House and had fought long and hard to make the land and buildings turn any kind of profit. I couldn't ever see him wanting to leave, no matter how tempting another farm might look. But we signed up to a couple of Scottish farm agencies and started to sift through the brochures they sent through.

Most of the farms were too expensive, with hundreds of acres of land. We could afford maybe two to three hundred acres with a farmhouse on site, but these types of farm steading were few and far between. Eventually we found two that we wanted to go and see. The first was just over the border into Scotland, and the second was on the west coast in Ayrshire. We chose a day when the kids were in school and made the long trip up the M6, past Glasgow and into the heart of the west Scottish countryside. Following the satnav was a doddle, but as we passed through one run-down town after another I started to feel uneasy. Eventually the directions led us through one of the worst villages I've ever seen: row upon row of grey pebble-dashed boxes with burnt-out cars in the streets and gardens littered with broken bikes and ripped-up sofas. It was like a battle zone, with sketchy-looking groups hanging out on street corners and feral kids roaring up and down the streets on quad bikes, with no helmets or parents in view. The farm we were driving to see was on a rise, just to the north of the town, and the farmhouse and yard were surrounded by tall security fencing with those lethal-looking, sharpened points to stop anyone climbing over the top. We pulled to a halt in front of a neat-looking farmhouse and stepped out of the car. The farm and the house were in good order, clean and tidy, and the buildings had no missing slates or broken windows so had obviously been looked after. Steve and I slowly looked around, and I turned to meet the agent, an earnest-looking man of around thirty who wore a green fleece waistcoat and a pair of expensive wellies.

After shaking hands, we walked across the farmyard, looking at the barbed-wire fences.

'The area seems a little… um… rundown perhaps?' I said, as Steve nodded in agreement behind me. 'Does the farm ever have any problems with theft or people trespassing?'

'Nooooo,' said the agent, shaking his head, 'there's nothing like that round here, no antisocial behaviour at all. In fact, there's been loads of government money spent recently and the new school is supposed to be fabulous.'

I thought about the building that we'd driven past on the way to the farm, a concrete block with tiny windows with a bright red security fence around the site. This bleak, grey block was obviously the school that the agent was talking about. 'In fact,' continued the agent, 'the area is pretty popular with tourists. The owners have a cracking bed and breakfast business.'

Steve raised his eyebrows at me as I screwed up my face. 'But... the fencing? It looks like they might be trying to stop people coming in?'

'Well, there is the odd one or two maybe,' said the agent slowly, 'but I don't think it's anything to worry about.'

Driving away from the farm, we stared straight ahead until we cleared the village limit and joined the motorway. Finally, I spoke: 'There is NO way on God's earth that I am ever sending our children to a village school like that.'

Steve nodded, his head bobbing up and down like a puppet on a string. 'Did you see the state of the town? Most of the streets had houses that looked like crack dens. Can you imagine trying to farm with that right next to you? I'd be constantly on the lookout for kids setting light to my haybales.'

'They seemed pretty convinced that it wouldn't be a problem,' I replied, staring out of the window at the grey drizzle spattering the glass. 'I mean the fencing is a bit of a giveaway. And did you see all the chains and padlocks on the buildings? And the CCTV? It's like Fort Knox in there, no wonder they're trying to sell up.'

Steve shook his head and drove on. We had another farm to see after lunch and it was on the east side, near the border of England and Scotland. As we drove the scenery started to change, and the grey pebble-dashed houses and bleak, flat landscape turned into miles of purple-tinged heather and pretty little villages made up of squat, red-brick cottages. Eventually we found the second farm, set back off the road in a green bowl of a valley that stretched into picturesque hills on the horizon. This farm was beautiful, with a handful of buildings built in white-painted sandstone and the coign stones picked out in black paint, all sheltering around a weed-free cobblestone courtyard.

The farmer, Colin, came out to meet us as we stepped out of the car. He was big and bluff, his face open and friendly with weathered cheeks, and he shook our hands and introduced his wife Susan, a friendly round-cheeked brunette, who stood in the farmhouse doorway wiping her hands on a tea towel. Colin showed Steve around the farm buildings whilst Susan invited me through the front door, stepping carefully over a huge, sleeping dog lying sprawled out in the hallway.

Susan showed me around her home, and I fell in love straight away. The building was Georgian and had the clean, straight lines and the high ceilings of the period. One of the rooms was circular with enormous sash windows and a pair of panelled curved cupboards set skilfully into the wall.

'They're original,' said Susan breezily. 'We've been told they were to keep plates and glasses in, but I use them for the kids' toys.'

All the floors were original, made from wide, oak planks with that wonderful deep brown patina that you only see when they've been polished over many, many years. Even though all the furnishings were beautiful there was a definite atmosphere in some of the rooms, a creeping chilliness that I could feel as I followed behind, shivering slightly and rubbing my arms to keep warm. Up a wide, dark staircase the second floor had huge bedrooms, with original wainscoting and massive windows, still with the original bumpy glass, brass doorplates and decorative plaster ceiling roses. These rooms seemed cold and bleak as the flat winter light fell in oblique angles through the large windows, bleaching the colour from the walls and curtains.

Susan paused on the landing and watched as I peered into each room, staring out of the windows at the garden beyond.

'We don't really want to move,' she said, her hand coming up to cover her mouth as if she knew she shouldn't be speaking.

I cocked an eyebrow. 'I thought your husband said you were wanting to move somewhere smaller?'

'Well, it's not our choice,' she said, looking down at the landing carpet. 'Colin's parents own the farm, and they want to sell up. To pay for their retirement, you see?'

'So you have to move out?' I said, watching as she blindly put her hand out to the side to hold onto the wooden bannister.

'Yep. But I don't want to, and neither does Colin.' She looked upset for a moment, but then continued in a rush. 'And the worst thing is that the sale is going to be split three ways, with his parents taking half, a quarter going to his stupid, bloody useless brother and a quarter to us. It's not going to be enough to find somewhere else.'

And with that sentence, she burst into tears, bringing her hands to her face as I stared at her in pity and horror.

'Oh my god. I'm so sorry,' I whispered, and I walked over to pick up her hand, which felt ice cold in my palm.

She turned away, wiping her tears on her sleeve as she waved me ahead of her to look at the rest of the house. 'No, go on, I'm so sorry. How embarrassing. It's just been rather a hard time.'

I quickly stuck my head into the rest of the rooms as Susan walked down the stairs into her kitchen. By the time I came downstairs, her smile was firmly pinned back onto her face and she was busily stacking dirty plates into the dishwasher.

Straightening up, she looked over at me. 'I am so sorry about that. I can't believe I was crying. Don't take any notice.' Just then Colin and Steve walked into the kitchen, cheeks flushed with cold as they chattered about hectares and land quality, stock numbers and yields. 'Alright pet?' said Colin, looking doubtfully at his wife and spotting her red and swollen eyes. Steve and I quickly made our excuses and walked back to the car, hand in hand.

'Oh man,' said Steve as we waved to Colin and Susan and pulled slowly out of the farmyard. 'That was bloody hard work. The poor man is being forced to sell by his parents. He was telling me all about it on the way up the track. He's worked there all his life, and now it's being sold so that his mum and dad and brother can live the life of Riley.'

I quickly told him about Susan's tears and how she was determined to pretend that nothing was wrong. 'The house was stunning,' I said, 'but you could just feel how sad it all was. Every room looked kind of mournful. And it felt really cold.' I shivered at the memory and wondered silently whether a house could pick up on the occupants' feelings and whether the bitterness and misery Susan must have felt at having to leave her home had crept into the walls and floors, leaching out any warmth and replacing it with a frigid chill.

Eventually the property was sold in lots, split up into blocks of land and the farm steading sold separately from the rest of the fields. It went for far more than we could afford, and in a way

I was glad as I wouldn't have wanted to be the reason for Susan and her family leaving the farm. I also wondered whether they had been able to find somewhere else and pick up the threads of a farming business again.

After the aborted trip to Scotland to look at farms, Brexit loomed on the horizon.

Steve and I had both voted Remain as we were both very worried about losing the European subsidies. They're essential to our small farm, and I couldn't see how we could manage without them. I also thought it was important that my children could live or work wherever they wanted in the EU, and after employing Nina in the brewery I could see how important workers from abroad were to local businesses. I also thought the current government lied through their teeth about how we'd all be better off after Brexit, and that bus and the strapline was bloody stupid, never mind being an outright fabrication. When the referendum result was announced I couldn't believe it, and I wondered how those in our industry who had voted for Brexit had felt that leaving Europe would be a positive development for UK farmers. However, people have their own opinions, and if working in farming has taught me anything it is that you can't force your own views down other people's throats.

It's a contentious issue, but in my experience, there are some farmers who have only ever worked for themselves, haven't had to travel for work and don't have a lot of familiarity with a multi-cultural society. The industry is mired in tradition, stubbornness and older ways of thinking. Of course, many farmers are forward thinkers and are forging ahead with new ideas and technology, but there are many others who hark back to the supposed 'glory days' of British nationalism with the accompanying suspicion of anyone who isn't of white, British ethnicity. The farming industry is still absolutely rife with racism and misogyny, and I wonder if there's any correlation between this and the fact that so many farmers voted to leave the EU. Of course many voted to leave the EU due to the perceived burden of bureaucratic red tape, but now that Britain is Brexited the red tape seems to have become worse, not better.

Leaving the EU was, in my opinion, one of the worst decisions that the United Kingdom has ever made. In one fell swoop we lost a lot of specialised, skilled workers, and the industry suffers from a shortage of trained labour. The double whammy of Covid and Brexit could be a death knell to many who make their living in the agricultural sector, from farmers to agricultural dealers and large animal vets to feed manufacturers. As ever, all we can do is try to ride it out and pray that all the different diversification projects we have set up so far help our small family farm to survive.

After Brexit happened, our chicken project idea was closed down by the bank and we came back from Scotland empty-handed; we had only one remaining opportunity left – to look at setting up a pig-fattening unit. A fortnight later we stood in an enormous pig shed in County Durham. It was so big I could barely see the far-end wall, and I watched as a thousand titchy piglets, separated into big straw pens, ran about squealing and playing or flopped into the straw and snoozed under massive heat lamps. We'd been invited to see another farmer's set up, and my immediate thought was: if the finances stack up this could actually work. Really, what could go wrong?

We had a contact in British Quality Pigs (BQP), a huge company that arranges pig breeding and fattening contracts with farms, and they had set up the visit. The farmer, Ralph, was in his late twenties and showed us the photos from his latest skiing holiday whilst we leant over the shiny, new-pig shed doors.

'So it makes a good income then?' asked Steve, peering into the shed.

'Aye, it helps,' said Ralph. 'Keeps the farm ticking over all year round.'

I reckoned if he could afford a week skiing in Croatia, then it was probably doing better than he was letting on.

Ralph let us into the shed, and we walked through the big straw pens, as he pointed out the different watering and feeding systems. The piglets were twelve weeks old, with long, heavy bodies and large flappy ears, and they peered myopically at us as we walked past. The smell was robust, a pungent concoction of salt and muck and straw. Most of the piglets were fast asleep, half buried under a thick layer of bedding with only their tiny noses poking out of the top, but some were ears deep in the pig feeders or playing in the wallowing hole at the end of their pen. I was fascinated by it all, soaking up information about different breeds of pigs, typical piglet behaviour and the recommendations on enrichment in a straw-based system. Ralph pointed out the automatic feeders and waterers and showed us how he checked over the piglets each morning, looking for any lameness or dehydration or signs that a smaller pig was being bullied by its peers.

As soon as the visit was over and I stepped back into the car, I turned to Steve to say, 'That's it! I think having pigs would be a fab idea. Did you see their tiny little faces?'

'Yes,' he responded tentatively, 'but they don't stay tiny…'

We chatted all the way home, trying to plan out where we could build the shed on our farm, how to persuade the bank to lend us the money and how Steve would manage all the extra work.

Back home, Steve started to pull a proposal together, and we began to work out the figures, creating spreadsheets to send to our bank manager containing proposed costs and our ideas of profit and loss. BQP were beyond helpful and came to visit High House to talk through our ideas and give us information about the different options.

We nervously sent off our proposal to our bank manager and waited for his reply. We waited. And then we waited a bit more. Chasing him didn't help as he told us that he was at the mercy of the 'higher-ups' and they were taking forever to look through all the different mortgage applications.

A full eight months after our application, our bank manager finally got back in touch.

'It's not positive, I'm afraid,' he said on the phone. Steve's shoulders slumped.

'I don't understand,' he said, his voice cracking slightly. 'You said that we'd definitely be accepted. And we sent you all the stuff you needed.'

The bank manager sounded embarrassed and a little defensive. 'I'm sorry, but it really wasn't my decision. They wouldn't rubber stamp the proposal as they said the risk was too high.'

'It took you *EIGHT MONTHS* to work this out?' said Steve, his voice rising in volume. 'Do you know how long it took me to put together the proposal? I thought you were supposed to be a friend to farmers. That's what all your crappy adverts say!'

I took the phone out of Steve's hand and spoke into the receiver. 'Right. Thanks for that,' I said down the phone.

'I'm sorry,' the bank manager stuttered, 'but the state of farming at the moment…'

I cut him off. 'Yeah. Thanks. Speak to you soon.'

I put the phone down and watched Steve as he sat down at the table, his head in his hands.

'Why is *everything* such a struggle?!' Steve shouted, looking completely deflated. 'Why do we never get a break? I mean, for fuck's sake, I've bent over backwards for that bloody bank, and they won't help us at all.'

I made him a cup of tea and then sat down across from him. 'Who else can we ask?' I said. 'This can't be the only place that we can go to for a mortgage?'

One of the main things I admire about Steve is his resilience in being able to pick himself up off the floor, dust himself down and carry on when many others would have given up. That afternoon he was on the phone to the Agricultural Mortgage Corporation, and they agreed to look at our proposal and give us a decision in a couple of weeks. That fortnight was torturous as we waited on tenterhooks to hear whether they would give us the go-ahead on the pig project.

'If we don't get the money,' I asked one morning after yet another sleepless night, 'what are we going to do?'

'We'd have to sell the farm,' said Steve. He sat up in bed and thumped his fist on the duvet cover in time to each word. 'This is my last fucking attempt to make it profitable.' He rubbed his eyes. 'If they say no, we'll put it up on the market, use the money to buy a house out in the sticks and I'll have to find a job in an office somewhere.'

After a fortnight our contact rang back and gave us the glorious news that they would fund the loan and we could start work on the planning and building works. The feeling of relief was immense, but it was slightly tempered when we received the contract and I realised how much money we'd need to get up and running. This was exacerbated when our solicitor called us in for an appointment, only to look me straight in the eyes and say, 'This mortgage is between you *and* Steve, you realise? So if he doesn't make the mortgage payments, then you will both be liable.' His voice rose on the last few words, and I squeaked a tiny 'yes' in reply, whilst I shakily signed my half of the loan documents. God knows how I'd make the payments if Steve popped his clogs and there was no one to farm – I'd have to sell up immediately.

We didn't waste any time in organising planning permission and arranging builders. At the bottom of our back field, a two hundred by fifty-foot pig shed slowly rose from the ground, built by an experienced team of builders from Suffolk who told me that they spent their life travelling across the UK, living in hotels and hostels and building steel sheds for new farming projects. Our shed was a thing of beauty, made from galvanised steel and concrete in an amazing fourteen days. The weather was unseasonably hot, and the builders swarmed over the pilings, dawn to dusk, shirts off and sweating in the sun, the light glinting off the steel and purlings. Then on a very bright September day we received a phone call from an exceedingly posh neighbour who was in direct eyeline of the new building.

'This shed,' he roared down the phone, 'what colour will those end gables be? They're so bright in the sun it's glaring right into my bloody orangery.' Steve placated him with the news that the gable end would be clad in an extremely expensive, non-reflective, forest-green, box profile sheet cladding. 'Jolly good!' clanged the Extremely Posh Neighbour. 'Glad to know you're thinking of your neighbours!'

We had a special contact at BPQ called Mark Jagger who I consistently, despite constant reminders from Steve, managed to call Mick at every opportunity. He sat patiently with us and explained how the pig unit would work. BQP had contracts with some of the biggest pork consumers in the UK, including Waitrose, Marks & Spencer, Morrisons and Greggs, and Mark explained that *our* piglets would eventually be destined for the shelves in the Co-op as part of their high-welfare, British pork range. The piglets weren't actually ours but were on loan, as it were, from BQP and we were just a small link in a very complicated and extensive chain of farmers who produced pork to the highest welfare standards in the world. To illustrate his point Mark carried around a scale model of a typical pig shed, made out of cardboard, and he slapped it down on top of our kitchen table.

'So this is the manure passage,' he said, pointing at the little cardboard corridors that ran the length of the entire shed. 'Pigs like to be clean, so they'll poo in these corridors to keep their straw beds nice and clean, and it makes it easy to scrape out the slurry with your telehandler.' The shed was designed with twenty long pens that stretched horizontally across the building, with open windows, automatic feeders and waterers at each end. 'And these pens can be adjusted,' said Mark, touching a finger to the cardboard model in front of him, 'so that when the piglets grow bigger they'll still have plenty of room.' To demonstrate he scrabbled in his backpack and brought out a handful of toy piglets made out of pink, squashy plastic. He plonked them down in the pretend pens to show where the piglets would be sleeping. The toy pigs were completely out of scale, towering over the walls of the model, looking like enormous, porcine dinosaurs, completely dwarfing the tiny, plastic figure of a farmer that had been glued to the plastic base.

'I hope the pigs aren't going to be that big,' I said, grinning at Steve. Mark rolled his eyes.

'So we don't actually *own* the piglets. They still belong to you guys, and we're just fattening them up for a few weeks?' I asked Mark, trying to appear sensible. He nodded his head.

'That's about it. They'll be with you for twenty-two weeks and will start being sent to market when they reach forty kilos.'

'D'you think there'll be at least one small, sad, special needs piglet that I can bring into my kitchen and love?' I asked hopefully, looking across the table at Steve.

He sighed and put his elbows on the tablecloth. 'No, because they don't belong to us. So you can't just adopt some pathetic thing and dress it in nappies, like you do with the lambs. They count all the piglets in and then count them out,' he explained.

Mark looked backwards and forwards between us. 'They probably wouldn't notice if I hid it in the cupboard under the stairs,' I muttered, glaring at my husband.

Mark cleared his throat. 'Okay, that's not going to happen as one, the piglets grow up really fast, and two, we have strict regulations on their welfare. You can't just adopt one and keep it as a pet and hope we won't notice.' I scowled and he carried on explaining how the automatic water and feeding systems would work, how we could add vitamins and minerals to the water and all about the automatic window coverings that came down to cover the shed in case of cold or windy weather.

'Pigs feel the cold, so you want to make sure that when they're very small they're kept warm and cosy. They tend to sleep on top of each other anyway, but there's a temperature sensor in the shed and you can check all the info on the computer,' said Mark.

I was impressed with the highly sophisticated system with essential facilities controlled from a computer programme that Steve could install on his laptop and his mobile phone. The piglets were to have toys too, to keep them occupied and stop them fighting with each other. Mark showed us pictures of the latest 'pig enrichment' products: long, steel chains that hung from the ceiling; hollow, plastic balls; and chunky, rubber chew toys that were strong enough to survive being chomped by hundreds of piggy jaws. 'But the best thing to keep them busy is lots of deep straw,' said Mark, 'and you can throw in cardboard, they love to chase that round, or even dig up squares of turf and chuck that into the pens. Pigs like to root and burrow and something like a big tree branch or even a handful of big pebbles will keep them occupied for hours.'

Steve and I listened as Mark explained how good stocking routines were the key to a successful fattening programme. 'Checking them twice a day isn't enough. You have to walk through each pen, clean out the water and see which one is lame or if there's a piglet that's being bullied by a pen mate. It takes a fair bit of time each day and even more when you factor in having to scrape out the muck.'

I chewed on the skin around a fingernail, already worried about how Steve would cope with the extra work. He was flat out as it was and was easily doing sixteen-hour days, seven days a week. He flashed me a smile and I bent to look at the model again, using a finger to move a huge toy pig along its pen.

Steve would do *anything* to keep the farm going, and I knew he couldn't wait to start properly on this project even though his workload would almost double in one fell swoop. I reminded myself that I'd try to broach the subject of hiring an apprentice for a few hours each week in an attempt to give him a bit of a break once in a while.

The next few weeks rolled past in a blur as we threw ourselves into the pigs, our time being taken up with usual planning tussles with Northern Electric, the local highways department and Northumbrian Water. The highways department was particularly awkward. We had posted them our plans and explained that we'd be creating a private track that would carry the big feed delivery lorries to the pig shed. After three or four weeks they eventually read our letter and gave me a ring.

The man on the end of the phone was to the point. 'You can't have a road as it doesn't have the right entrance onto the main roadway. You'll need a Category 3 entrance built, with raised gutters and road markings.' I listened in disbelief. What they were suggesting would cost an absolute fortune as it involved a proper tarmac surface, A-grade gutters and pavements and even an extensive drainage system.

'But the road running past our farm is only single track and it is *full* of potholes,' I replied, thinking in frustration of how little work had been done on the council-owned road since last year. 'You lot haven't exactly pulled your finger out to mend it.' Once a year the council sent a small van and an even smaller man who used a can of spray paint to mark the biggest holes and then came back six weeks later with a wheelbarrow and a spade to throw a bit of sloshy tarmac into the bottom of the craters. A few weeks of rain and the hole would appear again, the sticky tar floating down the road and taking a layer of topstone with it.

The man huffed on the other end of the phone and repeated his stipulation. 'You'll need to speak to the building department to find out what you need to do.' Eventually, we gave in and spent an extra £15,000 to install a sophisticated entranceway with black tarmac that ends abruptly when it reaches the council-owned road, the glossy surface butting up against the craters of the moonscape that runs past our house.

Finally, the shed was ready to receive its first delivery of a thousand piglets. We were all very nervous, but BQP sent us an angel in the disguise of a big, hairy Yorkshireman called Shane who had a blonde, bristly beard and tattoos all over his arms and legs. He reminded me of a (Yorkshire-accented) Viking, and he was a godsend as he had worked with pigs all his life and his straightforward, laid-back attitude was immensely reassuring.

Steve had decided that with the launch of the pig project, we should all be kitted out in a brand-new uniform in an attempt to make us look like a professional, skilled team. He'd bought a batch of crimson polo shirts and had them embroidered with our new High House Farm logo – a tree against a background of rolling fields. We all got one, the children and my mum and dad included, and I could see Shane blinking in amusement as he first caught sight of the entire family lined up against the front of the pig shed, wearing these bright red shirts and, for some reason, standing in height-descending order. Dad was especially pleased with his shirt and wore it with a pair of bright orange slacks and a pair of green wellies. Shane shook my dad's hand and, slowly running his eyes up and down his scorching new outfit, remarked, 'Bloody hell. You look just like a traffic light.'

We saw the pig delivery lorry drive onto our new road and watched it pull up in front of the shed. It was an enormous three-tier articulated truck with slats on all sides and had been loaded a few hours earlier from a free-range pig-breeding unit in Lancashire. As the enormous vehicle reversed onto the loading ramp, I peered through the little window gaps, trying to get a first glimpse of our new piglets. The driver pulled to a halt and stepped out, his hands full of different types of paperwork that needed to be filled in before the piglets could be unloaded. The piglets were five weeks old, had been weaned a few days prior and squeaked and squealed and rustled inside the truck.

'Okay,' said Shane, signing the last of the papers with a flourish, 'everyone ready?'

We stood just inside the shed with sweaty palms and nervous stomachs, waiting for the moment when the lorry doors would open. The driver lowered the ramp and made a few clucking noises and suddenly, out of the dark interior a *huge* throng of tiny piglets rushed down the slope and swarmed into the shed, squealing and grunting as they galloped over the concrete. They were so keen to leave the truck that they crowded the entrance, pushing and squeaking and climbing over each other to get to the front of the queue, one or two of them doing an actual forward roll as they all rushed down the ramp. I stood with my mouth hanging open as the hundreds and hundreds of boisterous piglets in shades of pale pink, taupe and chocolate brown, with long eyelashes and tiny, curled tails, bounced past me at high speed.

'They are the cutest things I have ever seen in my life,' I breathed, as the piglets trotted past, tiptoeing along the passageway as if they were each wearing a tiny pair of high heels. Some of the braver animals sniffed curiously at my waterproofs and wellies but skittered away in surprise when I bent down and tried to scratch their backs.

'Aye, they're not like pet pigs and handled or anything,' grunted Shane, 'so they're not very tame.' Shane had obviously seen many, many tiny piglets in his time and seemed absolutely immune to their innate cuteness.

We carefully herded groups of fifty piglets into each big pen, which had been bedded up with thick barley straw and roofed against any cold draughts. We had twenty huge pens, each with a shiny automatic water drinker, feeders and a raised bed to keep the piglets warm and dry. The piglets snuffled and squeaked among the straw, digging their heads right into the bedding and shoving their little pink noses at each other and at the pig toys hanging from the ceiling.

'Are you ready?' shouted Shane over the noise of a thousand piglets excitedly inspecting their brand-new home. Dad and I were to be the Vaccination Team, and we needed to inject every piglet to protect them against various ailments including the rather medieval-sounding pig-wasting disease. Shane had given us a huge box of vaccine vials, a brand-new automatic injector and a pair of weapon-grade earmuffs to stop us going deaf (or in Dad's case deafer) from all the screeching and squeaking.

'We'll have to sex them too and move them into separate pens,' said Shane.

He held up a squirming piglet, pulled up its tail and thrust its rear end at my face.

'What am I looking for?' I said, peering at the tiny pig's undercarriage. The piglet tipped back its head, opened its jaws and bellowed at the ceiling, making a deafening, high-pitched scream to protest against being picked up and having its bottom shown to a stranger.

Shane rolled his eyes. 'Its bollocks! It's got titchy bollocks!' he shouted, over the cross-screeching, waving his hand at the pig's bottom. 'Can't you see? This one is a boar and needs to go into the right-hand pen.' He gently placed the piglet over the wall into the neighbouring pen and stared at me with his eyebrows raised.

I honestly couldn't see what he was referring to as, apart from a tiny botty hole, each piglet looked very smooth under its tail like a little pink, plastic toy. I nodded vaguely, hoping that my expression transmitted an in-depth understanding of porcine genitals.

'And this one,' continued Shane, lifting up a titchy brown animal, 'is a girl! See?' He lifted its curly tail and waved its bottom in front of my face. 'So it's called a gilt and it goes into the left-hand pen.' The little piglet was duly deposited on the other side.

I was utterly confused, my eyebrows drawn down in bewilderment as I leant down to stare at the bottoms of various piglets that were scampering around the concrete floor.

'Oh my god,' huffed Shane, 'you better stay at the injection end, and I'll get t'other lads to do it instead.'

'Thank god for that,' I muttered and snapped my ear defenders over my head.

Dad held the vaccine vials and helped refill the syringes, whilst Mum and the kids shooed the piglets into the pens. The vaccination process was hard work: we were dusty and hot, and I couldn't believe the volume of noise the little pigs made when being scooped up by Steve or Shane. They *screamed* when they felt themselves being handled, and after a while, even wearing my defenders my ears were ringing with the screeching. As soon as they were put down each piglet immediately went silent and started nosing into the straw and investigating its new home.

Dad – struggling with his asthma in the dusty, straw-filled air – started to cough and splutter, and we stopped for five minutes so I could find him a face mask to protect against the swirling, gritty clouds. I was sweating so much it was running down my face and into my eyes, and it was hard to keep my mask in place as I pressed the automatic vaccine gun against each piglet's neck. The children were doing a brilliant job, chasing down any tiny escapees and gently shooing them back into their proper pen whilst my mum appeared with a huge picnic lunch, her arms groaning with different types of sandwiches, quiche and lots and lots of pork pies. We stopped to eat and sat outside in the shade of the roof, gratefully drinking gallons of water and tucking into the huge sandwiches. My arms and back ached, and everyone looked absolutely knackered. We'd been working for three solid hours and had still only injected about half of the piglets in the shed. Dad was grey with dust beside me but still full of enthusiasm, chatting away to Shane, sharing stories about his engineering days and roaring with laughter at his pithy replies.

The rest of the day passed in a blur of heat, dust and squealing piglets, until eventually every single little pig had been given the appropriate dose of vaccine. The pens started to settle down, the occupants draping themselves over each other in a warm huddle and shoving themselves bodily under the straw to fall asleep in big groups against the back wall. Steve bent over, his hands on his knees, and tried to stretch out his back, curving his spine against tight muscles. He was covered in scratches from sharp piglet hooves, red scores down his arms and shins, and he lifted his sweaty polo shirt to show me deep purpley scratches covering his stomach, where he had held each piglet close whilst I gave them the injection.

Shane didn't seem any the worse for wear, and after a quick rundown of points to remember when we did the late-night check, swung himself into his truck ready to drive to his hotel in

Newcastle. We waved as he drove off into the gathering dusk, his headlights shining over the stubble field on his way out to the road.

We still had to check the sheep, so Steve collected his quad bike and Mavis the dog and puttered off across the fields. The rest of us slowly walked back to the house, wincing against the occasional ache and pain and comparing scratches and bruises. Thank god we only had to vaccinate the piglets once, so we wouldn't be expected to do this again until this lot had gone to the abattoir and a new load had been delivered.

There was great excitement the next day when a huge lorry rocked up at the farm and deposited a battered shipping container next to the pig shed. Steve was drinking his morning cup of tea and grabbed my hand and pulled me to my feet.

'I've got an office!' he said, waving his arms like a windmill, tea sloshing out of his mug. 'With a desk and chair and everything!'

We walked down to the pig shed to watch the lorry driver carefully winch the container into place, and after he'd found the keys Steve flung back the battered metal door with a flourish. The container was painted a faded orange, had an enormous scrape down one side and smelt of mildew and dust.

'I found it cheap, secondhand on the web. It's been sitting in a scrapyard for years but it's got a carpet and a window and I'm going to connect it up to the electricity and the internet. The desk is coming tomorrow, I've ordered a kettle and I've even got a noticeboard to put up,' he said, waving his hand towards a large, rectangular package by his feet.

'Looks fab,' I said, my heart brimming over as Steve bent to admire a boxed-up mini fridge that stood waiting to be plugged in. 'You'll be able to move all the hundreds of farm files clogging up the desk in our house.' Steve rolled his eyes as he strode happily around his new Portakabin.

'What about a toilet?' I said. 'You're miles away from the house.'

'I'll go behind the wall,' he said, refusing to let me diminish his excitement, 'or join the pigs in the poo corridor.' I laughed and gave him a hug as he ushered me out of the container, carefully locking up behind him.

The rest of the day was spent with Shane, learning how to spot any problems with the piglets. Steve and I walked through the shed as he pointed out those that might have a problem.

'Look at that'un,' he said, pointing his finger at a small, greyish-pink pig that had huge, floppy ears half covering its face. I leant down to peer at the piglet as it snorted in alarm and tried to hide behind a bigger pen mate. 'It looks like Dumbo,' I said, nudging it gently with my hand. Dumbo sniffed my fingers with a tiny pink nose and bravely started to nibble on my trousers.

'Aye, it's the ears that let you know how they are. When they're down like that it's usually cos they don't feel well.' Dumbo flapped its ears and stretched, ambling over to Shane to nibble on the side of his boot. 'Just keep an eye on it, yeah? Sometimes they just don't thrive.'

That afternoon I drew back the bolt on the pen nearest the door and crept in, sinking down into the straw, my back to the high concrete wall. Immediately the crowd of curious piglets trotted over to investigate, nibbling my fingers and trouser legs as I put out my hand to scratch them along their backs. I noticed that they had surprisingly human-looking eyes with a coloured iris and white surround, and some were blue or even green or brown. Their tiny tails curved over their backs, and when they came closer and sniffed my hands the tails straightened out and then recurled in pure delight. When I touched their noses I discovered they were very sensitive but also very strong, and they seemed to use them like we do our hands, poking into all my pockets, lifting sections of hair or investigating my boiler suit, fascinated with the new textures and smells. One piglet, bored of my rubbery boots, nibbled the skin above my wellies, catching my leg with its needle-sharp tusks.

'Ow!' I squealed, flapping my hands in pain so that the whole herd of piglets turned and ran in one movement, diving into the straw at the back of the shed. After a moment I saw a few faces poke out and peer at me from a corner of the pen. Eventually, they began to creep back, the bravest ones in front, and resumed their snuffling and squeaking, fascinated by this new person who had entered their pen. It was interesting to see how different they were to our sheep and how quickly they became used to Steve and me and began to learn our voices and faces. One of the piglets would suddenly bark – a loud, abrupt, dog-like noise – and all the piglets would freeze, a foreleg raised as they sniffed the air with their noses. The bark was a warning noise, and when we first pulled back the doors every morning there was a chorus of similar deep yaps from the occupants inside until they realised it was just Steve and me coming to sort out their breakfast.

The new shed was designed with some very clever, automated shutters called Galebreakers that sensed when the weather was turning and would automatically close to protect the piglets from any high winds. Steve had been tinkering to set them up but hadn't managed to finish the job before the first autumn storm hit the farm.

It was a typical Northumbrian autumn day, with howling winds and horizontal rain, when I missed a call from Steve who was doing the morning checks in the pig shed. I trundled off to see what he wanted and found him in the middle of the pens surrounded by a noisy whirlwind of straw and dust. All the 'pig enrichment material' that we'd collected (including some gigantic cardboard boxes) was being blown around the shed followed by hundreds of very tiny and excitable piglets, careering around and squeaking in happy exhilaration. The high winds had also broken one of the gates, and it hung open, banging in the wind, as Steve swore his head off and tried to shoo the piglets into their proper pens. I watched in delight, laughing at the racing piglets as they careered from one side of the shed to the other, bouncing and jumping in joy. The chaos went on for quite a while, but eventually the piglets started to settle down, chomping contentedly on the piles of new boxes and snuffling after the shredded bits of cardboard had blown into the straw. After we'd tided everything up and closed the Galebreakers against the storm, I could still hear the occasional high-pitched squeak from an over-stimulated and still-excited piglet.

Every day the piglets grew bigger and bigger. You could almost see them growing; as they fed on the best-quality pig meal they quickly filled out, losing their piggy cuteness as their bodies grew longer and heavier.

My favourite thing to do was to replace the worn-out pig toys, and every week I'd collect up a huge armful of old Amazon boxes, bits of rope and felled branches and divide them out among the twenty pig pens. The piglets adored these new playthings and would charge around in excitement, skidding from one end of the shed to the other as they caught the cardboard in their jaws and flapped it around in the air. They would spend hours burying favourite bits of branch and shredding the rope, and alongside regular piles of clean straw they seemed to be happy to eat their fill, play until they were tired and then stretch out in the sunshine snoring huskily in unison. I occasionally sneaked a favourite pig a bit of vegetable or fruit, and it was a joy to see them carefully reach a questing snout over the barrier to gently hook an orange or whole cucumber and crunch it up in one bite, making messy lip-smacking noises as it enjoyed the treat.

A few weeks after the piglets arrived, I woke up one morning with a tummy full of what felt like rumbling lava and only just managed to reach the bathroom before the resulting explosion. The feeling didn't abate, and every hour or so I'd have to fling myself back on the toilet for another repeat performance. I took myself off to bed, lying pale and pathetic against the pillows.

'God, I hope it's not Covid,' said Steve as he inspected me a couple of hours later.

'I think it's typhus,' I croaked from beneath the duvet. 'I've been googling and I've got all the symptoms.'

Steve tried not to smile. 'Well let's do a Covid test and then see how you feel.' I dragged myself down to the kitchen and did an LFT test, shoving the little plastic tile across the table at Steve. Negative.

'I've just been on the phone to Shane,' he said, eyeing my dressing gown and fluffy slippers. 'He reckons it's the piggy shits. Everyone who gets pigs seems to get it, but the good news is it doesn't last that long, and once you've had it you can't get ill again.' He narrowed his eyes. 'Have you been washing your hands after being in the pig shed?'

I thought guiltily back to a few days ago when I'd helped muck out the shed, made my lunch and then spent thirty minutes cleaning the gunk out from under my fingernails with my teeth before I realised I hadn't washed my hands. Oh.

The piggy plague lasted four days, and I drifted unhappily in and out of sleep, damp tea towel on my head, surviving on cups of tea, Diet Coke and the occasional bag of salt and vinegar crisps. Steve had to work, although he did offer me an injection of Scabivax on the off-chance that that would help. I felt much better after a few days, and I was pleased to see that I'd lost half a stone in weight and relished the cries of 'oh you look so SKINNY' from my friends. 'You could always come over and lick a pig,' I said, as they bemoaned their own expanding waistlines.

As the pigs grew bigger, the smell of the pig shed became rather overpowering. We used to have a friend called Yoghurty Alan who worked on a dairy farm, and no matter how many times he washed or changed his clothes, he still smelt of slightly off milk. Steve was beginning to turn into the piggy version of Yoghurty Alan, and the thick, pungent smell seemed to follow him wherever he went. He started taking his overalls and wellies off before he reached the house, but sometimes the smell was so strong that it turned my stomach.

'Maybe you need a separate washing machine in your new office, or just some different clothes or something?' I suggested.

'Why don't I do the pigs whilst I'm naked?' snapped Steve in reply. 'I reckon they'd all like a good chew on my dangly bits.'

'Good idea. Then I could pressure wash you down before you step in the house.'

The pig slurry smelt revolting, but it was incredibly valuable to the farm. Full of potassium, nitrogen and phosphorous, it was the gold standard of organic manure, and local farmers gave it the nickname 'black gold'. We were keen to spread it on the stubble field next to the house before Steve drilled the next crop of spring barley, so we asked Jonty, our contractor, to visit with his muck spreader and dress the bare soil with the valuable fertiliser. Unfortunately, we had picked a very busy weekend for the glamping site and the brewery, and as Jonty's muck spreader trundled up and down our field, throwing clods of stinking muck out of the rear, the piggy aroma started to hang heavy and ominous in the air. The field was tight up against the house and the stench was so strong that our breakfast tasted of manure, and I had to bring in my washing from outside as the stink was permeating the clean clothes. Out of my window I could see a horde of glampers clustered at one end of the site, noses raised as they sniffed the air in bewilderment, so I ducked beneath the windowsill before they could spot me and start complaining about the disgusting reek.

I hid guiltily when the wedding party drove up and the bride, dressed in a beautiful gossamer lace gown with matching cream shoes, stepped out of the car and was immediately struck by an appalling whiff of pig shit. The pong was so concentrated it was like the odorous equivalent of a slap in the face and I could just imagine how it must have contaminated the bridal party's al fresco canapes and champagne as well as frizzling everyone's nose hairs. I have no idea if there were any complaints, as I deliberately sat in my room all day with the curtains closed against the smell and avoided all mentions of High House Farm on social media. Fortunately, later in the afternoon the wind shifted south, so although we could smell every particle of pig inside our sunroom and on our decking, the aroma eased off at the wedding site and glamping tents.

As they grew into adults the pigs became more boisterous, wrestling and fighting with each other, nipping ears and tails. They sometimes drew blood, and Steve and I were kept busy popping in and out of the pens to squirt a silvery aluminium wound spray on any bitten tails and ears. We kept careful watch for any weaker pigs that were being bullied and quickly transported any persecuted animals into Candy's old stable, bedding them up with thick straw so that they could recover from being harassed. Candy was not very impressed about her stall being used as a porcine hospital and would hide behind the bush at the bottom of her field in suspicion at the sound of oinking drifting across the stable door.

After sixteen weeks, all the piglets were enormous: fully grown with huge tusks and rippling muscles all over their long back and hind ends.

'If someone slipped in the pen and fell down, they'd eat them, wouldn't they?' I said to Steve as we leant over the upper gantry watching the huge animals snuffling in the straw below.

'Yup. Remember that serial killer that used to dispose of his bodies in his pig pens?' he said.

I shuddered. 'Well, if one day I go to check on the pigs and I don't come back, come and find me. I'm too small to fight back. There'd only be a pair of empty wellies left in the pen.'

'You're so titchy you'd be a piggy starter,' said Steve, ruffling my hair as we stared down into the pens.

After twenty-two weeks Shane decreed that the pigs were big enough to go to slaughter and started to organise the first of the pig lorry pick-ups to take our animals to the abattoir and meat-processing plant.

That evening Mark Jagger rang our landline and explained that we'd have to postpone this first pick-up, as (in his words) 'there was a bit of a balls-up in the abattoir.' Brexit (and Covid) meant that the industry was struggling to attract enough skilled processing staff, and the backlog of pigs needing to be slaughtered was rising every day. Our pigs would have to wait another couple of weeks before they could be collected.

'But they're ready now,' said Steve. 'Shane has graded them, and they need to go this week. Any longer and they'll be beyond the grading and I'll lose my bonus!'

Mark was furious. We were beside ourselves with the thought that our first experience of pig finishing was marred by Brexit, and the incompetency of the government and their failure to plan for skills gaps in essential industries. We started to hear about other pig farmers culling their pigs, being forced to kill them as young piglets and weaners, as they knew there would be no way there would be any space for them once they reached slaughter weight.

After a couple of weeks Shane managed to wangle a collection of one hundred and twenty of our biggest pigs, and each week after that we managed to send a paltry eighty or ninety animals to the abattoir. The pigs were now fully grown, heavy and aggressive with each other and us, and episodes of tail-biting and self-mutilation began to break out, in spite of the BQP vet trying her utmost to keep the animals calm. Steve and I watched in disbelief as Boris Johnson was interviewed on BBC News about the crisis in the pig industry, completely ignoring questions about pig culling and burbling on about a 'British Renaissance' instead.

Eventually, after fortnightly intervals of the pig lorry appearing for another animal collection, all of our pigs had made the journey of being loaded onto the vast lorry and transported directly to the abattoir, destined to be turned into high-quality welfare pork for the Co-op. I watched the last group of pigs trot up the ramp and felt a pang of sadness that they had experienced such a short life – it was such a different way of farming to our slow-grown lamb, born to pampered ewes that lived in luxury, each young animal allowed to mature over an entire year on the grassy fields on our farm.

I reminded myself that we had given our pigs the best-quality care possible, and the demand for pork (twenty-six thousand pigs per week are needed to make Greggs sausage rolls alone) meant that this was our best chance of making the farm profitable. The fact that we'd received our first cheque from BQP made a difference as well, and we started to ready ourselves for the next delivery of tiny piglets in two weeks' time. In that fortnight Shane organised a cleaning contractor for the shed, and they spent three

CHAPTER TEN

Our Future in Farming

Before I became a farmer's wife I knew it would be a life of hard work, but I hoped that I'd be able to settle into a way of rural living that would embody calmness and tranquillity. Compared to the life I led in the city, the farm is the epitome of slow and sustainable living, but living at High House has also challenged me in many ways that I couldn't foresee. I do enjoy long stretches of quiet peacefulness, but I've also been pushed out of my comfort zone in so many ways. Who would have thought I could lead tours of the brewery, become a wedding organiser, travel to book signings, lamb an enormous Texel sheep or muck out a thousand pigs on my own? Becoming a farmer's wife meant that I've also learnt to be a good businesswoman, to be constantly searching for the next avenue of income and the next way we can leverage the farm to protect it against future challenges.

Ever since the brewery and tearoom opened, we've had enquiries from visitors asking if they could camp on the farm. Over the years we've hosted caravanning clubs, motorhomes and many of those titchy little camper vans that seem to have just enough space for a sink and a toilet. Wedding guests were keen to pitch a tent, and brides and grooms have even spent their first married night under canvas at High House, usually setting up their tent in our paddock and spending the whole night being stared at by a very curious Scabby or being forced to share their breakfast with the fat pony. After a few years of hiring Portaloos and being forced to shovel up the occasional human jobby left behind by a hungover camper, it seemed sensible to think of expanding our camping offering and look at building a toilet and shower block instead.

Steve came back from checking the sheep one morning with a thoughtful expression on his face.

'I've just met some people who own a glamping company,' he said, swiping a bit of my toast on his way to the teapot.

'What did they want?' I asked.

'They call themselves Dark Sky Glamping, and they're renting a field up on Hadrian's Wall but they want to move. They were asking about renting our paddock instead. Nice couple. Kirsty and Jason. I think he's Glaswegian. I kept having to ask him to repeat everything.'

'Is it a good company?' I asked, poking my phone to google their current website. I scrolled down the page to see their photos and a plethora of glowing testimonials from previous clients. It certainly looked like a professional set-up, but I wasn't convinced that we should turn over yet another piece of our land for another tourist business. Our kitchen window looks right over the paddock, and we'd always stocked it with the more fragile (and moth-eaten) members of our sheep flock. It was nice to look out and see them bumbling happily around, snoozing among the tall grass or browsing the thick thistle patches that grew along the wall.

'I've asked them to come and meet us next weekend, to see if they're still interested,' said Steve, peering closely at our kitchen calendar and marking the date. 'Come and ask them a few questions and see what you think.'

The next weekend, over a cup of coffee and slices of carrot cake, Jason and Kirsty chatted to me about their business, laying out their brochures and business cards on our kitchen table. They certainly seemed enthusiastic about the idea of moving their bell tents down to High House and had plenty of ideas about ways to brand and market the site. They were very keen to work with the existing brewery and wedding business and told us how they'd done something very similar in East Africa, working with a partner business to hold successful wedding safari holidays and honeymoons. They became even more animated when we told them about our plans for a shower block and Jason asked us if we'd thought about adding an office or a shop.

'A shop?' I said, thinking of the brewery, which already sold casks of ale and bottled beer plus food and drinks.

'Yeah,' said Jason in his soft Scottish burr, 'to sell bread and milk to the glampers, and camping essentials. A lot of camping businesses have something on site to stop guests needing to visit other places. It might be a decent money earner if we pitch it right.'

After finishing off his slice of cake, Steve disappeared to show them around the site and where he was thinking about putting the new shower block. He came back flushed with success.

'They seem ideal. And renting out that paddock and the shower block will be a good way to earn a bit of extra income,' he said.

'Yet another project,' I sighed around a mouthful of carrot cake. 'I'll start the business plan for the bank loan tomorrow.' Glamping holidays were already very popular in the UK, aided in part by the strict Covid regulations and problems of travelling abroad. For once, the bank didn't have any problems with our plan, and within a couple of months the loan for the new shower block was agreed, planning approved, the contract signed, and Kirsty and Jason became part of the High House family.

Steve worked hard over the winter to lay down ten gravel tent pads and miles of winding pebble walkways in the paddock, all ready for the installation of the bell tents in the spring. Our builder started work on the shower block, and it slowly began to take shape, with four high-quality wet rooms, a disabled bathroom, titchy shop and even smaller office. Kirsty had loads of new ideas and every day she arrived with her car full of things to 'dress' the new site including potted trees and flowers, barbeques and fire pits and a play house and mud kitchen. She even set up a tiny library in an old, painted cabinet that she filled with secondhand books and games.

As the site began to take shape, we started to receive some weird phone calls on our home phone number. The phone would ring late at night, and when I picked it up I'd hear a muffled, male voice shouting and spitting swear words over the line.

'You can fucking fuck off with your fucking tents,' ran one shouty late-night conversation.

'Wha..?' I stuttered, straining to hear. It sounded as if the caller was standing in the middle of a very busy nightclub, with shrieks and shouts and a thumping baseline.

'I'll get an axe!' slurred the unknown man. 'I'll get on the tents and I'll fucking slice them up!'

'It's that man again,' I called to Steve, 'he sounds just… a little bit cross.'

Steve reached for the phone, but by the time he'd put it to his ear, the caller had clicked down the receiver and all he heard was the dialling tone.

The delightful nuisance caller started leaving threatening, expletive-filled messages on our answerphone, railing against our plans for a glamping site, shouting that 'he'd fucking get us all, fucking destroy the tents and caravans!' The police told us he was calling from an untraceable line and warned us to keep alert for any intruders or suspicious-looking activity.

This was the last thing we needed, with Steve up to his eyeballs in farm and building work as he was trying to do a lot of the site development himself to save as much money as possible.

At the same time, we noticed that a couple of roadside A boards advertising the glamping site kept being flattened by a heavy vehicle, such as a four-by-four or tractor. The original A boards were completely destroyed, but Kirsty and Jason replaced them and this time we padlocked them to the base of a telegraph pole. The same thing happened again; the padlock chain was smashed open and the signs were pulled apart and thrown into the nettles at the side of the road. A couple of weeks later, our joiner told Steve that he'd been enjoying a quiet pint in the local pub when he'd heard a heavy, thickset man laughing with his mates about how he'd 'fucking destroyed the signs' with his four-by-four and had 'cut the chain with a pair of bolt cutters'.

Steve recognised the description of the perpetrator and I couldn't understand the depth of this man's hatred. Why on earth would he be so upset about a small glamping site being set up on a back road on a nearby farm? Why start making nuisance calls to a local family and destroying signage? Why were some people who lived in the countryside so filled with bitterness and spite?

'It's always been like that,' sighed Steve. 'Some of them are either jealous of what you're trying to set up or have made up a weird backstory in their heads about how we have so much money and how their life is so much harder. Even though in reality we're absolutely skint. And they've never worked or lived anywhere else, so they're in this tiny echo chamber bubble and obsessed with local gossip. They have zero self-awareness. If they knew the actual truth…' He trailed off.

Before I lived with Steve, I'd imagined that he lived in a close-knit local community, filled with apple-cheeked neighbours who would be friendly and be happy to see thriving local businesses and who would all understand the pressures of running a rural business. Of course, there were many who were wonderfully supportive and always keen to see how we were diversifying our farm, but by god, on the other side of the coin there were those that bloody weren't and would like nothing better than to see everyone pulled down to their own low level.

Steve decided to take matters into his own hands and told the police our suspicions. They made a few quick visits and although no one admitted to anything, they told us they had 'an interesting chat with a very nervous man'. We also spent a morning reinstating yet another A board, and this time wrapped the security chain around a massive metal pole that was sunk into a solid chunk of concrete. A couple of days later we noticed that a car had taken a run at the sign, but instead of obliterating it like the last couple of times had only crunched into the concrete and left a minor dint behind.

We received word that a local was stuck at home as his car was in the repair shop; he'd hit something, and it had destroyed the front bumper and ripped off half the chassis underneath. Funnily enough, after that the phone calls stopped, and we've had no trouble since.

Over the next month I watched Kirsty put up her bell tents and start decorating the insides. Each tent had a double bed, fluffy rugs, wicker baskets full of extra blankets, matching cushions, a chair and table and a cute little ottoman full of kitchen equipment. Battery-operated fairy lights were threaded around the central pole, lanterns were hung from the roof and an electric fire was installed, with artificial flames to make it look like a real wood-burning stove. The effect was cosy and magical, and I could understand how this luxurious take on camping had become so popular.

Once the glamping site opened, we loved sitting out in the warm evenings, watching the fire pit flames flicker and listening to the quiet conversations between glampers or the squeal of happy children running about the site. Of course, there were the odd one or two that walked where they weren't supposed to, and I found the occasional startled glamper in the sheep shed, who had ignored all the warning and no-entry signs and had come face to face with some very interested sheep.

I started to host short farm tours, taking a family of glampers around our farm on a supervised visit to see our pet lambs, the pony, the tractor and anything else that I thought a visitor might be interested to see. After a while I grew more confident and started to weave in some fairly dreadful jokes and anecdotes about our pet lambs, the tups and the fat pony.

Whenever the fat pony saw me appear while shooing another gaggle of adults and small children into her paddock she trotted smartly in the other direction, determined not to join in. If I forced her to join in she'd either pretend to be dead and refuse to get up or, in complete contrast, start to crowd the smallest children, pushing her body into their faces and demanding tummy scratches and extra-strong mints. I'm convinced that she knew when a farm tour was about to start, as she often chose that exact moment to roll in the septic tank run-off in her field or start to gnaw on some rancid turnip that we'd left out for the tups. Eventually I created a small area hemmed in by an electric fence that I called my 'Meet and Greet paddock'. If a tour was booked, an hour before I'd catch a suspicious Candy and shove her in the paddock with a bucket of feed. She'd wolf it down then sulk in the corner, one hind foot cocked up behind, until I bribed her over with mints and tummy scratches to meet her adoring public. As soon as the farm tour was finished, I'd undo the electric fence and she'd slouch out and go to stand at the furthest point away from the footpath, pointedly scratching her bottom on the dry stone wall.

Most groups were really attentive and interested in the farm, but it was difficult to keep some visitors' attention on my tour.

They'd either wander off to poke around in the Barn of Mystery (so called as it's piled high with mysterious-looking, agricultural junk) or try to pick up a protesting (and poo-daubed) pet lamb. One morning I had a large group consisting of the world's most uninterested adults who ignored my jokes, stared in the wrong direction and interrupted all my good bits. It didn't help that Steve was cleaning out the dog kennel and kept talking over me every time he thought I'd made a mistake.

'So this is Mavis, our collie dog. She's six years old,' I said, waving my hand at Mavis who was lying in the sun with her tongue lolling out of the side of her mouth.

'She's seven!' yelled Steve from the back of the kennel.

'And she knows around twenty different commands,' I continued.

'More like thirty!' bellowed Steve in the darkness behind me.

I narrowed my eyes and closed the kennel door, deliberately shutting off his attempts to derail my talk.

'I'll show you a little bit of what Mavis can do,' I said, clicking my fingers at the collie, who bounced ahead of me as I walked to the front gate.

I ushered the tour group to stand just in the entrance and told Mavis to lie down. She dropped obediently to her belly, her eyes locked on the nearest huddle of ewes. The tour group looked fairly uninterested, and out of the corner of one eye I spotted that Steve had opened the kennel door and had stepped out into the farmyard, shading his eyes.

'So Mavis is trained to always keep on the opposite side of the sheep to the handler and will listen to my commands to see which way I want her to go. She's like lightning and very clever and is a brilliant working dog.'

I clicked my fingers at Mavis and shouted the 'Away!' command, expecting the dog to streak around to the right, running behind the flock to drive them up the field towards me.

Instead, she thumped back down on the ground, ears cocked quizzically and flicked her eyes back to Steve, who was now sitting on the tractor bumper with a big grin on his face.

'Away!' I shouted, a little more desperately this time, and the dog moved forward a few inches then turned around to sit by my feet, staring longingly back up the drive at a smirking Steve.

'Come on, Mavvy,' I crooned, feeling a flush creep across my skin, 'show us what you can do.'

Mavis yawned in response then got up and trotted out of the open gate, back up the yard to Steve. She plonked herself down beside his feet, her head turned up in adoration as he stroked her ears.

'Right. Well. That didn't go as planned,' I said, trying to laugh off the whole failed experiment. 'As you can see Mavis will only work for me if Steve isn't around to distract her.'

Steve wiped his hands on a rag and, taking pity on my embarrassment, clicked his fingers and quietly gave the 'Away!' command. Mavis bolted through the front gate, legs a blur as she shot around the entire flock, obediently gathering them up and driving them up to the gate.

'Lie doon!' shouted Steve and she dropped immediately to her belly, tongue lolling as he patted her head in thanks.

I stomped back up the farmyard, glaring at Steve, and I launched into the next section of my tour, making sure that we were as far away from Mavis and my husband as possible. That was the last time I asked Mavis to demonstrate her skills as I have a feeling she'd just disappear and show me up again.

The kids were always the best part of any tour, and I became used to toddlers hanging onto my welly boots as they saw the size of our CLAAS tractor, their mouths a perfect O shape when they realised that the wheels were at least four times their size. One little boy was amazed when he saw our ewes and lambs in the fields and ran over breathlessly to ask a corker of a question.

'How do them lambs get out of their mum?' he shouted in excitement.

I glanced at his parents and then explained quickly that when the baby was ready to be born, it came out of her bottom feet first and landed on the straw.

He stared at me for a bit, mulling the new information over.

'Well, how do they get in there in the first place?'

Fortunately, his dad stepped forward and distracted him by pointing out our quad bike, prompting him to rush over and demand to be lifted onto the seat.

The glamping site was a huge hit and more families started to book up weekends and over the school holidays. Kirsty had the brilliant idea of setting up a 'Fairy Trail' in our ten-acre wood and handed me a box full of tiny, plastic fairy doors that she'd bought at a discount on eBay. 'You can write the story to go with it,' she said brightly as I clutched the shoebox to my chest. 'You're good at making stuff up.'

Once in the wood, I stared around the tangled interior trying to decide where to put my first door. We always left the ten acres of woodland almost completely untouched, felling the occasional dangerous dead tree to let a little light onto the floor of the forest, but leaving everything else to grow how it wanted. The forest had two tiny streams running through, and it was thick with brambles and fern, studded with elderberry and holly bushes and boasted an expanding population of foxes, badgers and birds. Entering under the trees felt like walking from daylight into night, as the background noise of the wind and the calling of birds suddenly stopped as you stepped into the hushed, green-tinted twilight of a typical English woodland. I love our little wood and find it so peaceful, although it can sometimes feel a little haunted and watchful, especially in the wintertime when all the greenery has disappeared, and all that is left are branches and bare earth. My children don't like playing in it at all, and sometimes complain of 'feeling watched' if they do venture into its depths. It's easy to get lost in the middle and I've often found myself heading in the wrong direction if I don't recognise any landmarks.

Interestingly, very recently I was searching the British newspaper archives online and came across a sad article from 1962 about a plane crash on our farm. Two 'piston-engined Provost planes' took off from nearby Ouston Airfield for a night training flight and collided in mid-air. One plane smashed into a turnip field and one into a cornfield. Both student pilots were killed, one man trapped in his burning plane and the other managing to bale out but his parachute failing to open. He was killed on impact, and as the newspaper reported, his body was 'flung into the wood and found in a tree'. I've never managed to find any more information to pinpoint the exact location of the crash, so if any of my readers have heard the story and know any more details I'd be interested to hear from you. (Note: The story was covered in the *Liverpool Echo* on Tuesday 14 August 1962.)

Today the wood felt stuffy and hot as the summer sun beat down on the leaves outside. It was also full of humming insects, and I slapped irritably at the back of my neck as little harvest flies started biting my exposed skin. My friend Rob had come over to help, bringing with him his Labrador Diva, a screwdriver and some titchy screws to fix the doors to the base of the trees.

I pulled out the first door, running my fingers over the moulded plastic and pulling the tiny doorknob with the tip of my fingers so I could open the brightly coloured door and look through the empty gap.

Rob took it from me and pointed out a likely-looking tree, carefully positioning the door at the base of the trunk and starting to screw the tiny surround to it.

'These little screws shouldn't hurt the tree,' he grunted, as he niftily wielded the screwdriver. Diva sat at his feet, panting heavily in the warm sunlight.

'I've always been a bit disappointed by the concept of fairy doors,' I said, as I sat down on a nearby branch. 'It's so exciting seeing them on the tree, but then when you open them up, there's just trunk behind. I was always hoping for something like Narnia.'

Rob sat back on his haunches, admiring his placement of the bright pink and green plastic door. 'That's life though, isn't it?' he replied. 'You think there's excitement and magic, but in reality we're grownups with responsibilities.' He patted Diva on the head and walked on, spotting another tree beside the path.

I sat down in a patch of sun, watching Diva snuffle among the brambles as Rob worked on, his screwdriver squeaking as he fixed all eleven fairy doors to trees, most tucked away at the base of their trunks, but a few higher up, sitting on branches or holes in the bark.

'D'you think these are too high?' I said, inspecting a door that Rob had installed three feet off the ground and at the back of the trunk so that it couldn't be spotted on the footpath. 'Will anyone find it?'

'Don't worry!' said Rob. 'Families want a bit of a challenge. It's no fun if you can just find them all straight away. It'll be a great test to see if the kids can hunt them out.' (Note: I should point out that Rob is a big, tough ex-soldier and enjoys stuff like triathlons and twenty-five-mile yomps in the pouring rain. For fun. And therefore hasn't much experience of setting up one-mile fairy trails.) We picked up the tools and walked to the house, chatting as we wandered back over the front field.

Writing the fairy story didn't go as straightforwardly as I'd hoped. I sat staring at the screen of my laptop in disgust, waiting for inspiration to strike.

'I'm no good at writing tinkly fairy stories,' I said to Steve, groaning as I tried to write a few sugary lines about some twee woodland fairies. I struggled on for a few minutes then put my head on the desk in defeat. 'I can't do this, dribbling on about Flossy the Fairy.' I tapped my biro on the dining-room table in thought. 'I know, I'll tell them that if they find a fairy ring and leave some Marlboro Lights in the middle, they'll be guaranteed a visit from that sexy blacksmith from the next village along.'

Steve stared at me silently over the top of his *Farmer's Weekly*.

'You're no fun,' I said defiantly. 'If you don't give me some ideas, I'm going to write a story about Bill the forest fairy who has a twenty-a-day fag habit and a gammy leg.'

With Steve's help, my twee story featuring cutesy gnomes and quaint fairies was finished and I sent it off to Kirsty to be turned into a proper leaflet. Kirsty reported that whilst her visitors enjoyed the story no one could find all the fairy doors, and they usually only managed to spot a paltry seven or eight out of the full total of eleven.

'Told you!' I said triumphantly to Rob. 'We've made it too difficult, and no one can find them all.' We trooped down to the wood in an attempt to reposition the more difficult-to-find doors. Except we couldn't bloody find them. After half an hour of searching through fronds of bracken and staring at tree trunks we'd only managed to find seven and we couldn't remember where we'd put the rest.

'Oh for god's sake,' I said. 'This is ridiculous. I definitely remember counting eleven.'

Rob rubbed the back of his neck in frustration as he stared into the matted brambles. 'I definitely remember bolting them to the trees,' he muttered as we both walked deeper into the middle of the wood, circling each oak or ash tree to see if we could spot a door on the other side. After another spell of searching we gave up, and the upshot is that I still don't know if someone has nicked them or whether they were hidden too well by an overenthusiastic Rob. Let me know if you ever manage to spot the full eleven.

One morning, I was hanging out in our garden and spotted a skinny teenager, dressed in tracksuit bottoms and expensive trainers, skulk down the farm driveway, giving our machinery barn a suspicious sideways glance as he walked past. He was followed by a gaggle of similar-looking teenagers, desultorily kicking a football between them as they rounded the corner and went out of sight into the front field.

'I'm keeping an eye on those boys,' I said to Steve as I watched over my garden gate. He shaded his eyes against the sun, spotting the group as they opened the gate into the sheep field.

'Don't be so harsh on the kids,' he said, 'they're probably just having a game of footie or finding a quiet place for a fag or something. Can't you remember being fifteen and on holiday with your parents?'

I didn't reply, remembering that at fifteen I had been a huge geek and spent most holidays happily playing Monopoly with Dad or reading Terry Pratchett novels on my own. 'They're not kids, they're *youths*,' I replied darkly, 'and look like they're up to something.'

'Oh my god,' said Steve, 'you're like a smaller version of Miss Trunchbull.' I elbowed him sharply and went back to peering down the farm drive, my hands shading my eyes so I could see against the bright sunlight.

When I went round to check on the pet lambs, I spotted the same group of teenagers holding something in front of their tent and laughing as they chucked it at one another and tried to duck out of the way. As I walked over, they all started to look nervous and one of the lads ducked into the tent awning, carrying whatever it was out of sight.

'What've you got there?' I called across the fence, glaring at them as the boys moved from one foot to another. A tiny blonde girl, obviously a younger sister, wandered up to stroke a pet lamb.

'They've been finding stuff in the barn,' she piped up, ignoring the frantic shushing noises from her brothers behind her.

'Have they?' I said. 'What did they find?'

'Nuthin',' said the ringleader, whilst the adorable blonde tot looked up into my face. She frowned.

'He did! Kieran found the big knife thing and he's got it in the tent,' she added, pointing into the interior of the bell whilst Kieran sent her death stares in reply.

'Hand it over,' I said to the boys. 'Come on, it's probably dangerous.'

Kieran huffed in disgust and disappeared into the gloom of the tent, reappearing with an enormous medieval-looking scythe with a sharpened blade towering high above his head. 'I wasn't going to do anything with it,' he mumbled as he handed it across the fence.

'This is lethal!' I squeaked in reply, feeling the edge of the blade with my finger. 'We use it to chop down nettles in the sheep paddock.' The boys all hung their heads in embarrassment as their tiny blonde sibling looked at them in triumph.

'I telt them,' she said in a small, firm voice. 'I telt them that they'd get told off if they nicked it.'

'Stay. Out. Of. My. Sheds,' I enunciated clearly at the boys. 'There's stuff in there that's very, very dangerous. Don't go into any of the farm sheds. Didn't you read all the warning signs?'

Kieran shrugged and turned away. 'Didn't see anything,' he muttered in reply and stalked off into the distance, the back of his neck burning red in embarrassment.

'I'll let you know if he pinches anything else,' whispered the tiny sister conspiratorially. 'He doesn't listen to me mam or dad.'

Apart from Kieran and his light-fingered compatriots, we had very little trouble with the glamping guests and enjoyed meeting families from different parts of the UK. The glamping site was set up just after the first lockdown and everyone seemed to be so bloody grateful to be out of their houses and on holiday. You could see the relief in the parents' faces as their kids played outside in the sunshine, exploring the footpath and woods. It was wonderful to see families bonding over their glamping adventures, groups of friends sitting around their campfires chatting late into the night or kids squealing with excitement at their first sight of a lamb or a snoozing piglet. It reminded me very much of the exhilaration and delight I found on my first visits to the farm, marvelling at the beautiful scenery and getting up close to the cows and sheep and stroking Red Cow and Lop Ear all those years ago.

As the popularity of the glamping site grew and people began to share photos on Instagram and reviews on TripAdvisor, Steve and I started to receive requests for us to be in various farming campaigns, such as the NFU's 'Back British Farming' and 'Love Lamb Week'. Organisers usually only wanted a photo of me holding a poster or a simple quote, but in 2021 I was contacted by the communication team from Northumberland County Council who wanted to know whether we'd be keen to part of their new

'Love It Like It's Yours' campaign. The campaign was designed to welcome visitors to the county, but also to promote a few small steps to encourage tourists to stick to the Countryside Code. The advert would feature characters plucked from various industries across Northumberland who would each deliver a different slogan to the camera, on various topics such as keeping dogs on a lead, shutting gates and picking up litter. We were asked to be the face of 'safe parking', trying to encourage people away from parking in gateways and blocking a farmer's access.

Steve and I were happy to be involved, especially as we were well versed in problems with gorilla gateway parkers. Each year, usually during the summer, drivers would spot our large grassy gateways and decide they were the perfect place to pull up for a picnic, to walk the dog, or for more… ahem… dodgy shenanigans, not realising that they were in the way of a mad, sleep-deprived farmer who needed to cut twenty-one acres of wheat before the weather broke. On warm summer nights, Steve often stays out late, working in his tractor in the fields, only to find himself blocked in a gateway by a car with steamed-up windows and two half-naked and mortified adults trying to get their pants back on whilst being blinded by the tractor headlights.

The day arrived and the film crew appeared headed by an intense Art Director from a Newcastle-based marketing agency. They looked around the farm with interest and explained how they would film us in front of a field gate as we spoke a few prepared words.

Steve parked the tractor behind the gate, and we got into position – arms around each other, with me standing on a small rock so that it looked as if we were approximately the same height.

'Right Sally!' shouted the Art Director. 'When you're ready just read the words on the card.' A young assistant held up a piece of cardboard with the words 'Love It Like It's Yours' scrawled across in Sharpie pen.

I took a deep breath and opened my mouth, realising that my top lip was stuck to my teeth with nerves.

'Love it like… ' I started croakily and immediately ground to a halt as the Director flapped his arms and shouted 'Cut!'

'Let's try that again,' he said kindly. 'Try to smile this time and look like you're enjoying yourself.'

I stood up a bit straighter and plastered a pained grimace onto my face.

'Right,' said the Director doubtfully, 'off you go.'

I pushed back my shoulders and bellowed, 'Love it like it's yours!' at the camera, finishing the phrase with a rictus grin.

'Okayyyy. Can you do it in less of a posh voice?' said the Director apologetically. 'It just sounds like you come from anywhere but round here.'

I was starting to panic, sweat beading on my hairline as I tried to smile, stand up straight and attempt a thick Geordie accent.

'Love it like it's yourrrrrrrrs!' I squawked at the camera, trying to give the whole sentence a Newcastle-inspired lilt. Steve's shoulders started to shake, the ridiculousness of the situation and my appalling attempt at a Geordie accent making him snort with waves of suppressed laughter.

There was silence. Even the assistant holding the sign looked disappointed.

'Let's try Steve instead… ' suggested the Director.

Steve had a couple of goes at the phrase and the Director seemed satisfied. I sighed as the cameraman produced a high-tech-looking camera attached to a drone. I'd seen these before and knew that we'd be asked to walk across the field for a long panning shot, which would eventually be cut down to a microsecond during editing. We got into position and trailed across the field, trying to walk normally, avoid tripping over and being followed by a curious Scabby who wanted to know a) why we were fannying around in her field at this time of day and b) where was her dinner?

You can see the final advert online – Steve and I are on screen for about half a second. Amazingly, they used the take with my attempt at a Geordie accent, and in the panning shot you can also see me trying to surreptitiously shoo Scabby the sheep away from my coat pockets. We were also part of a social media campaign, and I carefully retweeted and shared every post from the marketing agency, often adding my own twist to their message, such as 'Please stop parking in our field entrances for illicit shenanigans with your other half and leaving random articles of stained underwear hanging from our gateposts.' Or 'Please don't do a poo in a plastic bag and leave it beside our gate, whilst throwing your used toilet paper into the hedge where it'll be discovered by our sheep who will enjoy the festoons of tissue as an unexpected afternoon snack.'

After a while the marketing agency didn't tag me into their posts any more and I settled back into anonymity. If any rich marketing agencies are reading this, Steve and I are always available and now have experience in reading straplines in dodgy Geordie accents, walking stiffly across a grass field and bickering gently out of earshot of the cameraman.

*

After twelve years Kirsty and Jason decided that they would like to expand and under the banner of 'High House Barn', took over the tearoom, wedding business and brewery in our buildings alongside their glamping business.

Kirsty asked if I'd help with the first few weddings to ensure all ran smoothly. I thought back to the early days, of the struggle to find good staff, problems with drunk wedding guests and the sheer strain of working on your feet for sixteen hours a day.

'I'll do what I can,' I said nervously, 'although I'm always anxious and need to be in bed no later than ten o'clock at night.'

'Ohhkayyy,' she said, opening her eyes wide. 'Well, we'll both learn together and between us I'm sure we'll cope with it all.'

And so far we have. The whole farm is now under one point of management and all the events, weddings, birthday parties, Christmas lunches and celebrations have gone smoothly so far.

And so we continue, with sheep, pigs, glamping, farm tours, weddings, lambing and harvest, juggling the myriad of businesses we've set up to try to make this little corner of Northumberland viable as a small family farm.

Looking back at the years before the farm I realise now that I couldn't have stayed in the corporate world. I now understand that I was fundamentally unsuited to that way of earning a living and utterly ill-equipped for the alpha-loving, materialistic world of the city. We did have some good times, the long lunches, boozy evenings and expense cards come to mind, but the breakneck speed and sheer noisiness of the lifestyle had an impact on my mental and physical health. I'm proud I managed to extricate myself in time.

Working in one of the 'big firms' meant that you were always chasing the next contract, the next client, the next bonus, and filling your life with the latest *things* – clothes, cars and phones. Success was measured by how much money you made for the firm and the investors. I felt trapped on the treadmill, making profits to fill the pockets of people that I didn't know, spending hours sitting in a tiny cubicle, staring at a computer screen and wilting under the flicker of fluorescent lights.

Farming attracts those who are patient and careful hard workers, rather than the hard-shelled extrovert prized by the corporate world: practical and innovative men and women who can draw on centuries of experience and create a thriving, healthy farm that produces something that is needed, at a *fundamental* level, by people.

It's not plain sailing of course. It's never that. So a sense of humour helps when things go wrong, and it feels like the god of farming is against you. Having resilience and the ability to pick yourself up and carry on is perhaps the most important trait in a successful farmer.

And farming hasn't been the miraculous panacea to *all* my worries... It is one of the most unpredictable industries in the world, and being at the mercy of government policies, poor weather and environmental issues means that I still spend my life with a constant low bubble of anxiety flowing underneath all that we do.

Oh god, and the lack of money! The sheer terror of not having enough cash and trying to work out whether we can continue to bring up our family at High House. The farm, in truth, is both a joy and a millstone around our necks. Owners of small farms are in similar predicaments all over the UK – all the money you make goes towards the upkeep of the steading, and there is very little to spare. We are sitting on a valuable asset, but the land and the buildings need constant work and don't bring in an income on their own. The grinding worry about finances is *relentless*, and we just don't dare think more than a couple of years ahead. I just hope that there will be enough so we can give our children what they need to set up their own lives in the future.

And sadly, I'm not at all sure that their future will be at High House. The state of the industry, the amount of sheer hard work and the poor returns mean that we hope that they'll find another career first before deciding if they want to take over the farm; they need to find a career in something they love (but also not squeeze themselves into a well-paid job because they think they should, as their mother did).

I am grateful every day that I was able to step away from working in a city and live in one of the most beautiful places in the world. I feel lucky to be able to graft alongside my husband and family in the open air, to look after animals that are pleased to see me (most of the time) and to provide food for people. Living in this way and following the well-worn groove of the agricultural year feels like a blessing, a very old and hallowed way of life that I know would be recognisable to the farmer living here one hundred, two hundred or even five hundred years ago.

We're hoping that in around twenty-five years we'll have managed to pay off our gigantic mortgage and overdraft and may even be able to retire. At the moment I'm still knee deep in all the gubbins of each farm project, and I can't picture how Steve and I will cope with retirement. I have a feeling that by the time I'm able to retire, I'll be down to around three foot high with two artificial knees. I know that Steve will struggle with not having a farm to run. I hope that he'll fill the gap by becoming intensely interested in something absorbing and mildly eccentric such as tinkering with miniature steam engines or making furniture out of leftover pallets.

I'm trying to take each day as it comes, not look too far ahead and concentrate on keeping my children happy, warm and fed. I do hope that in time our farm will be a little less fragile and a little more profitable, and with luck will be a thriving rural business which can be passed on either to new owners or the next generation. There are not many of us small farmers left – our numbers are shrinking all the time – and I feel I owe it to myself to enjoy and make the most of this opportunity.

*

On a bright spring morning, the children and I pull open the door to the lambing shed and take in a deep breath of air scented with the comforting smell of warm straw and sheep wool. The newborn lambs are bleating hungrily, and we can hear the deep chuckles of the ewes as they heave themselves to their feet and start shouting for their breakfast. Ben and Lucy collect buckets to fill with water as I busy myself with checking round each pen, making sure every lamb can stand up and has a full tummy. Ben climbs into the last pen and settles down with a tiny triplet on his lap – I hand him a bottle full of milk and he expertly pushes it into the lamb's mouth. Its tail starts to wiggle as it sucks, the little nose pointing at the ceiling as Ben gently pushes the teat into the hungry mouth. Lucy is right beside me, doling out buckets of water and pointing out any weaker lambs as she marks them down on the whiteboard at the top of the shed. She has a great 'stock eye' and is invaluable for spotting an animal that is off-colour or unwell. Steve joins us in the shed, shooing a new mum and a pair of babies into a smaller pen as he chats about his shift and how many lambs were born overnight. After an hour of work it's time for a hot drink, and we fire up the water boiler at the top of the shed. The children open a new packet of chocolate biscuits, and we all wander out to sit on a straw bale, hands wrapped around our cups of steaming tea or coffee, and munch companionably as we enjoy the warmth of the sun. I look around at our family: Lucy and Ben chattering about the latest lamb arrivals, Steve answering their questions patiently as he sips his coffee whilst the steam curls into the sky. I settle back and enjoy this moment of stillness, my heart full, surrounded by my family. This is what I was searching for, and this is what life should be about: the love for my family, our animals and our farm. And despite all the chaos, hard times, unexpected disasters and money problems, I wouldn't give this lifestyle up for anything.

All those years ago, when I first moved away from my city-centre flat and into Steve's tiny farm cottage, I was eagerly and nervously anticipating how life would turn out and how I could become part of the story and make my mark on this small family farm. And I can honestly say that I knew then, and still know now, that leaving the city for a life in the countryside was one of the best decisions I've ever made —no matter what farming misadventures that life might bring.

A LETTER FROM SALLY

Thank you for reading my book! If you enjoyed it and want to discover more inspiring memoirs, just sign up at the following link. Your email address will never be shared, and you can unsubscribe at any time.

www.thread-books.com/sign-up

What the Flock! was written in snatched moments, the quiet times before I rushed outside to lift bales of hay, feed a poorly pet lamb, muck out squealing piglets or charge into the kitchen to throw a dinner together for the kids. I think the hardest part over the last year was writing and trying to home school both children during a very busy lambing season whilst being under lockdown and in the middle of the pandemic. (Roughly around the second month of home schooling I threw out grammar quizzes and online maths lessons, and instead the kids learned how to attach the seed drill to the tractor and use the baler.) I think we *all* need a prize for pulling through the pandemic.

Lots of people have asked me whether they're in the book. And I always laugh uncomfortably and say 'No, of course not.' But even if they are, they're heavily disguised as I don't want to become a pariah in the village coffee shop or have rocks thrown at me during the summer fete. (Likewise, everything in the book *did* happen, but the timelines might be squashed or stretched a bit to make it flow better.)

People have asked why this book focuses so much on farmers' wives rather than farmers' husbands or partners. I suppose that the book reflects my reality, and even though there is a very welcome move towards increasing numbers of female farmers, the industry is still mired in tradition and heavily dominated by male tenants and owners. Sorry female farmers and their partners/husbands – I know you're out there and I hope the numbers increase.

I wanted to honour all the farmers' partners who, over many generations, have been expected to juggle *everything*: having babies, dealing with stressed-out husbands, looking after pet lambs/piglets/calves, giving a hand during the harvest, driving a tractor to move bales, looking after children, feeding family and farmhands, slogging away at endless housework and, ultimately, trying to keep body and soul together on a less-than-healthy income.

I always think about the hundreds of years of history at High House Farm. Our four-bedroomed house was once a one-roomed cottage with one single fireplace, a beaten earth floor and an outside water pump. Looking at the census, in the latter half of the nineteenth century it was home to a 'carter and horseman' and his wife plus six children under the age of twelve. I wonder how this family managed, with so many mouths to feed, cold winters, a lack of money and the constant threat of disease. But then maybe some things *do* stay the same?

When I check the sheep in the front field, the deeply worn track bears witness to the hundreds of shepherds who have walked round this field before me, checking their flock morning and night, carefully keeping an eye out for foot scald or fly strike. I love this. It reminds me of the hundreds of years of good husbandry and care that have led up to Steve and I working this land. On cold winter mornings, when it seems too cold to go outside, I motivate myself by imagining a long-forgotten farm worker hauling themselves out of a similarly warm bed to deal with a pregnant ewe or newborn calf. If they could do it, so can I.

In bad times, the history of the farm is a way to put our worries in perspective, and it's comforting to know that Steve and I are just a tiny link in a very long chain of farmers stretching back to Roman times, who have all attempted to scratch a living from High House Farm. I wish I could talk to them and learn how they managed and what the fields and buildings looked like in their day.

I hope that this book is helpful and gives encouragement to those who want to try moving to or working on a farm or a small holding. If nothing else, I've been as honest as possible and tried to give an accurate portrait of the problems and joys of living on a small family farm. Living in the countryside isn't for everyone, but I found living a slower pace of life and being outside every day calmed my overactive brain and suited my highly strung personality. And I've thrived… most of the time.

At the bottom of this letter, I've shared details of some of the most useful resources that are available to those in the farming community who might be struggling. If you are feeling under mental or financial pressure, please reach out and talk to someone. We've had help from the Royal Agricultural Benevolent Institution ourselves: they were incredibly helpful, tactful and discreet. And if you have a spare few pounds, please think of donating to them – they do amazing work.

One last thing! If you have enjoyed reading this book, I would be very grateful if you could leave a review. And you can always get in touch with me via Twitter (@pintsizedfarmer), Instagram (@pintsizedfarmer), Facebook (@pintsizedfarmer) or my website, sallyurwin.com. I would genuinely be delighted to hear from you.

Again, thank you so much for reading this book. Here's to many more adventures at High House Farm!

The Royal Agricultural Benevolent Institution (RABI) are a national charity that provides local support to the farming community across England and Wales. They have provided guidance, financial support and practical care to farming people of all ages for generations.
www.rabi.org.uk

The Farming Help partnership is a collaboration between The Addington Fund, The Farming Community Network and Forage Aid and is supported by The Prince's Countryside Fund. The Farming Help charities are working together to support the farming community during the current Coronavirus (Covid-19) outbreak.
www.farminghelp.co.uk

The Farming Community Network (FCN) is the first port of call for pastoral and practical support for anyone in the farming community, with a particular focus on farmers and farming families. The helpline is available every day of the year, from 7 a.m. to 11 p.m.

www.fcn.org.uk

The YANA Project provides confidential support, mental health awareness and funding for counselling for those in farming and rural trades in Norfolk, Suffolk and Worcestershire and has compiled, published and funded a directory of regional support groups and key national charities that can specifically help those in the rural communities.

www.yanahelp.org

The DPJ Foundation was set up in July 2016 in Pembrokeshire, supporting those in the agricultural sector. The mental health charity has grown and covers the whole of Wales with all areas of support.

www.thedpjfoundation.co.uk

Mind provides advice and support to empower anyone experiencing a mental health problem. The charity campaigns to improve services, raise awareness and promote understanding.

www.mind.org.uk

GET CURIOUS.

Join our community

www.thread-books.com/sign-up

for special offers, exclusive content,
competitions and much more!

Follow us for the latest news

@Threadbooks

/Threadbooks

@Threadbooks_

/Threadbooks

LISS

Braccia

June

Judy

Aimee B

Tracy
Hawkins
J Walsh
Erica
O'Connell Blr
Padjoc?
Sha - Roe
Reddin B&B
Regina
Arthur

Made in United States
North Haven, CT
15 August 2022